ORGANIC FARMING:
YESTERDAY'S AND TOMORROW'S AGRICULTURE

ORGANIC FARMING:
YESTERDAY'S AND TOMORROW'S AGRICULTURE
by the editors of Organic Gardening and Farming® magazine

Edited by Ray Wolf

RODALE PRESS
Emmaus PA

Printed on recycled paper

Library of Congress Cataloging in Publication Data

Main entry under title:

Organic farming.

   Includes index.
   1. Organic farming. I. Wolf, Ray. II. Organic
gardening and farming.
S605.5.0675   631.5′8   77-5674
ISBN 0-87857-175-2

4 6 8 10 9 7 5 3

Contents

Introduction

What is Organic Farming?

Organic farming is both an agricultural philosophy and a farm management system. Contrary to what a lot of advertising and university research would have people believe, organic farming is not just throwing a bunch of seeds on a field and coming back later in the year to harvest what's left. Organic farming certainly is not going back 50 years in agricultural research, quite the contrary. Organic farming is using the very latest in technology, applied to current research, to fulfill the principles of good soil husbandry our forefathers adhered to.

Organic farmers are aware of the damage done to soil by applications of harsh chemicals. They believe that in order for agriculture to remain productive, the soil must be cared for, not abused. Organic farmers not only wish to avoid the use of chemical fertilizers and pesticides that can cause damage to soil and wildlife, and create toxic side effects in a variety of ways but they also are very much concerned about the prevention of erosion, the adding of humus and other organic matter to soil to improve fertility, the preservation of small family farms, localized marketing of food, energy conservation, and proper nutrition. It is a rare organic grower who does not share those concerns, or pursue those activities.

American farmland has been farmed chemically for no more than 40–50 years at most, and already many fields are listed as worn out, unproductive, and ruined. However, parts of China have been farmed for over 4,000 years, and the soil remains fertile and productive. The question is why?

The best possible answer comes from F. H. King, and the work he did in the early 1900s. King was chief of the Division of Soil Management of the U.S. Department of Agriculture and author of four books on agriculture. In 1906 and 1907, Dr. King traveled to the Orient and made the most careful observation of Chinese, Korean, and Japanese agriculture that had been recorded by an American up to that time. He was extremely impressed by the careful handling of organic materials by all Oriental farmers—a direct contrast to the wasteful and destructive methods of many American farmers.

The Oriental system, said King, produced a permanent, highly productive system of agriculture, with little erosion and small need for fertilizer. While U.S. farmers had already spoiled much land in Dr. King's time, Chinese and Japanese farmers were seen by him to be growing rice and other crops on land that had been in continual use for 4,000 years, and was still productive. Dr. King titled his book about that trip *Farmers of Forty Centuries*, and it is still being read and used by teachers and scientists today.

What Dr. King found so important to Chinese and Japanese farmers, is equally important to organic farmers in America today: the careful handling and preservation of organic matter. The adding of organic matter to soil to produce humus is the one overriding concept all organic farmers hold to.

Humus creates a biological process in the soil which releases minerals that are locked up in unavailable forms in the mineral fraction of the soil. These locked-up minerals are present in almost all soils in enormous amounts. If minerals are released slowly, and if the humus used in crop production is continually replaced, a soil can produce good crops permanently.

Synthetic fertilizers can easily create imbalances in the soil, and in the plants grown there. Nitrogen has been the cheapest and most widely used synthetic fertilizer. It stimulates lush plant growth. Phosphorus is the next-most-widely-used ingredient in artificial fertilizers, and also has the ability to boost yields. Organic farmers choose not to use these two items in their highly soluble chemical form. We at Rodale believe—and this view is shared by many agronomists—that those two major nutrients are so heavily used on conventional farms that the roots of plants are inundated with them, and for a variety of reasons lose their ability to forage for such vitally needed nutrients as zinc, calcium, selenium, and others. The result is the production of plants that contain mineral imbalances, or outright deficiencies.

Particularly in the area of needed trace elements, synthetic fertilizers can have a detrimental effect on food quality. The major impact of synthetic fertilizers is to provide the primary nutrients—nitrogen, phosphorus, and potash—which boost yields and apparently reward the farmer with the greatest financial return. Nitrogen in particular has been used heavily because it increases yields at seemingly low cost. However, introducing high amounts of nitrogen into the soil does have the potential to create a lush growth of plants with lower nutritional value. Organic nitrogen fertilizers are usually not concentrated enough to cause that effect.

Many of these mineral deficiencies are well recognized by farmers, which is why the feeding of mineral supplements to cattle and other farm animals is now an almost universal practice. However, the mineral nutrition problem is now of such magnitude that deficiencies in major crops should not and cannot be ignored much longer.

There are in the United States today vast amounts of raw or waste organic materials for composting and mulching. The statement has often been made by critics that "organic farming works fine, but there isn't enough organic material to go around." On the contrary, there is more organic waste around than we know what to do with. The handling and "disposal," if you will, of organic wastes is a great national problem, and many millions of tons of potential humus are going to waste, or are creating severe environmental problems.

Organic farmers do not limit their soil-building

activity to the use of compostable materials and manure. They also grow green manure crops. The term "green manure" covers a wide variety of plants, very often legumes (which fix nitrogen), that grow rapidly and provide large amounts of organic matter that can be plowed or tilled into the soil, and then allowed to decay into humus. Use of such green manures protects the soil from erosion while also adding fertility to the soil.

A wide variety of natural fertilizer materials is also used by organic growers. Limestone and pulverized phosphate rock are commonly used, as well as other natural minerals, and organic products of concentrated nutrient value, like cottonseed meal, alfalfa meal, animal bone meal, dried blood, hoof and horn meal, spent mushroom compost, fish emulsion, and so forth.

The insect and plant disease-control practices of organic growers also differ greatly from conventional practices. First, organic growers frequently choose to grow types of plants and specific varieties of plants that are more resistant to insects and disease than others. Many such varieties are available, and massive efforts are now being made by plant breeders to produce varieties that contain elements of partial and even total resistence to insects within their germ plasm.

By growing a mixture of plants in relatively small plots, and by avoiding use of insecticides that kill beneficial as well as harmful insects, organic growers frequently have large populations of predator insects and other organisms on their land. By that means, the population of harmful insects is reduced, and the damage they cause minimized. The strip-cropping also helps reduce erosion, and enables a wider variety of green manures and small grains to be grown.

Literally hundreds of biological control techniques are available, and are often used by organic growers. They range from companion planting and interplanting of two different crops in the same field, to the introduction of organisms that are pathogenic to target insects. Rapid strides are now being made in biological control, and it is the opinion of many entomologists that these methods, long favored by organic growers, will be more economic and certainly more safe in the long run than old-style use of highly toxic and expensive pesticides, to which most insects are able to gain immunity after several generations.

Organic growers, and also their customers, tend to be less interested in the total absence of signs of insect infestation on fruits and vegetables than are conventional farmers and the average supermarket customers. Frequently, to gain that total absence of signs of insect's presence on crops, a large number of both dangerous and expensive pesticide applications must be made. There is now a strong movement among both entomologists and consumerists to educate consumers to accept fruits, vegetables, and other foods that are perfectly sanitary and nutritionally equal, but which may contain signs of insect damage.

Organic farming is an agriculture of permanence. A method of crop production that aims to work as much as possible with nature, and accepts some of the restraints nature puts on us. Organic farming is aimed at promoting plant, animal, and human health, through the growing of crops on soil that is nutritionally balanced, not loaded down with the three major nutrients in a highly soluble form. And lastly, organic farming is profitable. Chapter 2 discusses the most in-depth study done to date on the economics of organic farming and shows that organic farmers spend less money than chemical farmers, but take in a little less money from the sale of their crops, and thus both groups end up making the same amount of money per acre a year.

What we have tried to do in this book is present the many facets of organic farming to familiarize readers with the complexities of this form of agriculture. After talking to those people involved in organic farming, and seeing their deep reverence for caring for our farmland, I can only say this is a book about the art of organic farming, as practiced by some of America's best farmers.

Ray Wolf

Correcting Some Misconceptions About Organic Farming

There are many misconceptions about what organic farming really is. To help clear up some of these errors Jerome Belanger, a long-time organic farmer and publisher of Countryside magazine, takes a critical look at the nine most often discussed myths about organic farming.

Myth Number One: Plants don't really know or care in what form their nutrition is delivered.

Even if this were true (and there is increasing evidence that it is not) it could still be included as a myth, because it begs the question. The form of the nutrients provided has a greater effect on the soil than on the plant. In particular, the effects of toxic chemicals on soil life affect the soil and consequently the plants.

7

A fertile, healthy soil may contain as much as eleven tons of biological life per acre; not only earthworms and insects, but bacteria, fungi, nematodes and actinomycetes. It is significant to note that a conventional farm using toxic chemicals can have as little as two tons of soil life per acre. This means that the application of two to three hundred pounds of chemical fertilizers, plus lesser amounts of various pesticides, has destroyed 18,000 pounds of soil life on each acre.

Because of the ecological networks which connect everything to everything else, the destruction of this protoplasm has consequences which reach man. For one thing, those nine missing tons per acre of life excreted waste, reproduced, and died in their subterranean environment, obviously contributing immeasurably to the soil and to plant life even if we only consider the manure value of their waste and carcasses. But in addition, the symbiosis between various classes of soil life and between that life and plants and the soil itself, is well established. Then, too, there is an ecological axiom that a more diverse population is a more stable population, and stability brings a resistance to disease and other problems. The diversity and stability of soil life is adversely affected by toxic chemicals.

Myth Number Two: If all farmers farmed organically, we would all starve to death.

This myth is tied to the mistaken notion that organic farming is simply farming without chemicals.

The fact is, many organic farmers are harvesting yields just as respectable as their chemicalized neighbors, and yields that are far more respectable in terms of food value, energy use, and care of the soil. This is most important if man intends to continue eating in the future, because organic farmers are not destroying our soil.

This myth can only be used by those who are not aware that there are already thousands of farmers tending organic crops and harvesting yields that compare favorably with chemicalized crops. And they are doing it more cheaply, using less petroleum which is the basis for nearly all agriculture chemicals, including fertilizers; they are doing it with fewer diseases and pest problems; and they are leaving the land in better condition than they found it.

Myth Number Three: There just isn't enough manure to fertilize crops to produce the food we need today, and if we went back to draft animals they'd require so much land for feed that we couldn't grow enough food to support the human race on the remainder.

This is another myth that demonstrates a basic misunderstanding of organic agriculture.

Manure is an important and valuable soil nutrient. Much (some authorities say most) manure is wasted, because of improper management. Too much is treated as waste, something to be gotten rid of, rather than as a valuable resource. Too much is lost through mishandling due to the lack of knowledge about the real role of manure in soil fertility.

However, manure is not absolutely necessary for organic farming, and in any event it certainly is not a complete fertilizer for most soils.

Organic farming is possible without either manure or compost, except for the organic matter of crop residues. Soil amendments are usually necessary in any event because few soils are "complete" for the crops we grow. Because the basic foundation of any soil is the rock it evolved from, and because rocks have various mineral compositions depending upon their origin, it follows that any given soil will be deficient in one or more of the more than 100 elements

9

found in fertile soil. These amendments are often applied in the form of rock dusts from areas rich in the deficient mineral.

Perhaps the best example of natural fertilization methods involves nitrogen; the most common, the most important, the most expensive, and the most damaging nutrient in the forms used by chemical farmers. Nitrogen is applied as some form of ammonia, such as anhydrous ammonia. This ammonia is especially detrimental to the very forms of soil life that produce nitrogen naturally, which means that the more nitrogen a chemical farmer uses, the more he needs.

Nitrification, or the production of abundant atmospheric nitrogen through natural means, provides this valuable and expensive plant nutrient at no cost. The legumes are most famous for nitrification, but there are nitrifying bacteria that operate without legumes. In fact, there are several products on the market now for treating seeds which are said to increase the nitrifying bacteria on nonlegumes such as wheat and corn. (This type of research and its acceptance by organic farmers is one more indication that organic farming is not "going back" to horse and buggy methods. In several areas organic technology is actually ahead of conventional agricultural research, but it is distinguished by its respect for natural laws and the interrelationship of all living things.)

Anhydrous ammonia, which is produced from petroleum, is 82 percent nitrogen. While it was first used only in 1947, today nearly every acre of corn is treated with it. Anhydrous is extremely caustic and is applied under rigid safety conditions—for the applicator, but not for soil life. Skin contact can cause frostbite and severe burns, eye contact with it can cause blindness, and inhalation can cause death. It must be banded deeper and farther away from crop roots than any other chemical fertilizer to avoid severe injury.

Given these properties it should be easy to imagine what its effect on soil life would be. In one case where anhydrous was applied through an irrigation system, it was reported that thousands of earthworms came to the surface and died within minutes. Most corn belt soil requires less than 100 pounds of nitrogen per acre, but that amount in this form kills tons of soil life, including the nitrifying bacteria.

An additional disadvantage is that anhydrous is in such a free form that most is lost into the air unless soil moisture conditions are just right. And none of the chemical forms of nitrogen are long-lasting; they are easily washed out of the soil to become major factors in water pollution. Excess nitrogen also affects the color, flavor, food value, and health of the crop.

In contrast, organic forms of nitrogen are more stable and become available for plant growth more slowly than the shot-in-the-arm forms. Organic matter holds nitrogen in the form of protein compounds, which give it up slowly under the pressure of microbial dissolution. The protein first decomposes into its amino acids, and then into ammonium compounds. These are broken down into nitrites by bacteria, and then into nitrates, which is the form of nitrogen plant roots can take in. It is virtually impossible for too much nitrate to be released at one time; it is released throughout the growing season instead of all at once.

Under the natural process, the more fertile the soil is, the more fertile it becomes. With the chemical process, the living soil becomes a dead thing.

Manure is extremely valuable in the fertilization process because of the organic matter and nutrients it contains, but it is not an essential component of organic farming except insofar as much of this material has been wasted in the over-reliance on toxic chemicals.

Myth Number Four: The soil is just a medium to hold the roots while you feed the plant.

This is an essential tenet of toxic technology because under those conditions the soil is virtually dead; it *is* just a medium to hold the roots because it has so little to offer the plants.

But a healthy, balanced, organic soil is alive. Because the webs of ecology are so complex and the variety of soil life so diverse, no one can say for certain what symbiotic relationships might exist between plants and even the tiniest, most insignificant form of soil life. Therefore a complete and balanced soil is the first requirement for complete and healthy plant growth. The organic farmer feeds the soil and the soil feeds the plant.

Baker and Cook, in *Biological Control of Plant Pathogens*, explain how bacteria, fungi, and actinomycetes all work together. "Bacteria are generally effective as scavengers and are thus important in competition; a few species may also produce antibiotics. Actinomycetes are poor as scavengers and in competition, but are excellent antibiotic producers. Fungi are effective in competition, hyperparasitism, and some produce antibiotics. Bacteria are effective in the rhizosphere (the zone of soil immediately surrounding plant roots) and all three are effective on the organic crop debris or crop residues during fallow periods."

Some species of nematodes (worm-like creatures parasitic in plants and animals but free-living in the soil) attack plants, but these are preyed upon by some fungi. Conventional farming methods, however, call for fungicides which destroy the beneficial as well as the less desirable fungi and therefore affect the nematodes. Dudington and Wyborn have shown that these fungi occur most commonly in soil that is high in organic matter. In fact, as far back as 1938, Linford and associates found that when organic matter is added to soil, the populations of beneficial fungi and free-living nematodes, but not the plant-eating nematodes, increase rapidly. The fungi and free-living nematodes then prey on the parasitic forms. This indicates that when fungicides are used to protect plants, more fungicides must be used in subsequent applications, and the battle will never be won. Conversely, if no fungicides are used and other cultural practices to encourage beneficient soil life are followed, no fungicides will be needed.

According to a 1969 study by Bowen and Rovira, bacteria "often occur as colonies which coalesce to form an almost continuous cover several bacteria deep on plant roots. Such bacteria may stimulate plant growth."

Feed the soil and the soil will feed the plant. The soil is not just a medium to hold the plants' roots, but a living organism that is closely integrated with the growth and health of the plants.

Myth Number Five: Organic farming simply means farming without chemicals.

Organic farming involves an entire management system which includes tillage practices, crop rotations, diversification, and even water management and wildlife protection.

In tillage, for example, the moldboard plow is the favored implement on most conventional farms. As a tool which quickly and easily breaks up the land to prepare a suitable seedbed, the moldboard plow has been acclaimed as one of the greatest contributions to civilization.

Hay is an important crop to the organic farmer. It not only allows a field to build up certain mineral levels, it also keeps weeds clipped preventing weed seed formation.

However, this plow (and especially the modern versions of it which are pulled by heavy and powerful tractors) has several drawbacks. It tends to bury the trash (crop residues) beneath the zone of the highest biological activity, thus putting it out of the reach of these microorganisms. At the same time, subsoil is brought to the surface, and this subsoil is deficient in organic matter, so that the microorganisms have nothing to work with.

The moldboard also creates what is known as a plow sole layer. As the bottom, or sole, scrapes through the soil, the effect is very similar to that of a cement mason's trowel, especially on heavy soils and soils that are worked when they are too wet. In time, this plow sole layer becomes impervious to water, to plant roots, and even to further plowing.

One alternative is the chisel plow. This implement looks like the curved fingers of a hand being drawn through the soil. The tines dig in and mix up the soil but leave the trash in the top layer where it belongs. Because the tines are relatively pointed and widely spaced, a plow sole layer does not develop. Continued use of the chisel results in better soil tilth or friability, which means better water retention with obvious benefits for both soil and water conservation as well as plant growth. An added benefit reported by farmers is that, as soil texture improves, plowing becomes easier, requiring less time and fuel. Farm-size rotary tillers are available but they can be used only in certain situations. For example, they require land that is rock-free. Chisel plows are actually less troublesome than moldboards on rocky ground.

Crop rotation is an important management practice on most organic farms in terms of both fertility and pest control. Corn, for example, is known as a heavy feeder: it demands fertile soil, and especially, nitrogen. If corn follows alfalfa or some other nitrogen-fixing crop in the rotation, the nitrogen comes from the air rather than from our dwindling supply of petroleum and natural gas. Similarly, land that grows corn year after year is often highly susceptible to corn rootworms which chew through the roots and destroy the crop. Insecticide attachments for rootworm control are standard equipment on most corn planters today, but crop rotation helps to prevent the pest from building up to dangerous numbers.

Crop rotations require a certain amount of diversity in contrast with the monoculture so common in conventional farming today, and that diversity also brings with it other benefits. It has been said that the usual "rotation" in the corn belt is corn, soybeans, and a trip to Florida. No livestock is kept, which not only means that no manure is available, but also that hay and pasture, or even such small grains as oats, are not needed or wanted and therefore crop rotation is not possible. Monoculture, hundreds and thousands of acres planted to a single crop, also affects the biosphere and therefore the ecology by limiting the environment and food for birds, insects, and other living creatures, all of which are related in one way or another to crops and therefore to man. Monoculture also helps to produce a hospitable environment for plant pests and diseases, and aids in their distribution.

Crop rotation does have other benefits. It is the management tool most organic farmers use as part of their weed-control program, for example. Weeds that gain a foothold in drilled crops such as oats, wheat, and barley, can become a serious problem without using herbicides, but an alternative is to rotate the field to a row crop such as corn or soybeans where cultivation during critical periods will help eradicate those weeds, or at least keep them under control.

Wildlife is important to the organic farmer, again because of the webs of ecology. Nearly everyone is familiar with Rachel Carson's *Silent Spring*, the book which first brought ecological awareness to the public attention and which held that pesticides used to kill insects also killed the birds that ate the insects with the result that there were fewer birds. It followed that more insecticides were needed because the reproductive rate of insects is greater than that of birds. Observing further that the most harmful species of insects become resistant to insecticides, the organic farmer applies this lesson to every other form of life, including the all-important soil life, and manages his activities accordingly.

Therefore, while the conventional farmer concentrates on overkill, painting himself into an ever-smaller corner, the organic farmer strives for natural balance in the ecosystem.

Myth Number Six: Making all farms organic would be a step backwards, because one farmer would no longer be able to feed fifty other people as he does today.

This again is based on the erroneous assumption that organic farming means turning back the clock. Organic farming is a technical and highly sophisticated system that bears little resemblance to the farming methods of the North American pioneers. The fact is, there are a great many modern farms of conventional size and respectable output using organic methods . . . and doing very well.

Myth Number Seven: Organic farmers are against technology and progress.

Research into organic farming methods is actually ahead of conventional agricultural research in some areas. As one example, the University of California, Riverside, recently "discovered" the value of mycorrhiza. The presence of this fungi in the root environment enables plants to take up greater quantities of nitrogen, phosphorus, potassium, manganese, and others. While the new data comes from conventional research, Dr. William Albrecht, a leading proponent of organic farming, published the same information forty years ago.

There are many other examples, including recent research in Iowa on foliar feeding, which has long been used by some organic farmers and ridiculed by the people who are now "discovering" it. Organic farmers are already working with a product which makes use of natural antibodies instead of antibiotics (and faces outright hostility from the establishment); bacteria that aid in nitrification in nonlegumes, and electronic seed treatment.

Organic farmers are not against technology and they definitely are not against progress. They are against the misuse of technology.

Myth Number Eight: Organic farmers are nothing but faddists.

Organic farmers and the people who support them are keenly aware of the limitations of our planet, and the ecological web which makes every living creature dependent in some way on every other. They have respect for rational science, while those who compartmentalize each specific area, who look only at this plant, that nutrient, or that bug, without connecting it to the whole, are guilty of dangerous tunnel vision.

"Fad" is defined as a practice or interest followed for a time with exaggerated zeal. By this measure, which is the fad: blithely splashing on fertilizers, squashing bugs, and stomping weeds with chemicals (the most widely used of which are less than twenty-five years old); or endeavoring to follow the tried-and-true methods of Nature herself, which have been the norm for millions of years?

Myth Number Nine: Chemicals are too entrenched to give way to organic methods.

It would probably surprise some staunch supporters of chemical agriculture to learn that organic practices are becoming common even on conventional farms. The cases where "respectable" researchers are discovering what organic farming researchers knew long ago is one example. The fact that farm-implement sales show that chisel plows are becoming increasingly popular and that many conventional agronomists are recommending the chisel for the same reasons organic farmers have long been using it, is another.

There are many more, some dramatic, some quite subtle, that are readily apparent to anyone who knows something about organic methods and has been reading farm publications for any length of time. These publications now abound with such scattered gems as veterinarians saying that perhaps antibiotics should not be used routinely and indiscriminately after all; economists speaking of "optimum yields" instead of maximum yields; and even conventional farmers gloating with justifiable pride over an increase in earthworms in their soil due to cultural practices.

Perhaps even more significantly (although on a somewhat grimmer note), university and governmental advice on such topics as manure management is becoming increasingly more common, and more "organic." A sharp turnabout was seen during the oil crisis, when fertilizer prices skyrocketed. Most people are aware that the respite in the oil situation is only temporary, and that chemical farming just will not be competitive in the future.

One of the most positive forces which is changing chemical farmers into organic farmers is the health problem related to chemicals, with which many farmers are faced. When a doctor tells a farmer to quit using chemicals or get out of farming, because of a skin condition or other problem, he may take a different attitude toward ecological methods.

But the greatest change of all is in the grass-roots mass movement against chemicalized food. Farmers are in business, and want to remain in business, and their interest in toxic and mutagenic chemicals is economic. If the public outcry against agricultural chemicals continues and grows, farmers will change in spite of the companies with heavy financial stakes in agricultural chemicals. The list of chemicals that have already been banned in the United States and Canada is long and growing, and still the news media picture dead fish and birds as well as other tangible evidence of the hazards of chemicals. Meanwhile, more and more information about their carcinogenic and mutagenic properties are reaching the average reader, and an ever-growing awareness of ecology in general is helping the common man put all the pieces together.

Organic farming is the wave of the future. It must be, for with continued and increased dependence on agricultural chemicals, there might *not* be a future.

Jerome Belanger
Reprinted from Alternatives, *Vol. 5, Nos. 3 & 4*

The Quiet Farmers Agribusiness Ignores

It's easy to drive right past Dick Lessig's little farm in Hancock County, Ohio, and not notice it at all. Like thousands of other small farms, there's nothing exceptional about Lessig's place by the standards we use to measure "progress" and "status." At first meeting him, you'd say there's nothing exceptional about the man, either. Together, they're just a plain, simple, 100-acre farm, a traditional old white barn and farmhouse, a line of old farm equipment, and a farmer more concerned with his team of horses than the price of corn in Chicago.

Agribusiness does not think Dick Lessig really exists. The champions of "get big or get out of farming" have predicted his demise for so many years that they have had to go ahead and bury the coffin— empty. Indeed, the number of small farmers is probably rising now, though there is no way to tell, because agribusiness, unable to predict the stubborn creature out of existence, has deftly tried to define him into nonentity. Only farms with high gross income and "full-time" operators are real farms, according to USDA criteria.

"It's how much you net, not how much you gross that matters," says Lessig, smiling as usual. "And if you check it out, you'll find that many of the big farmers are only part-time farmers, too." (I checked with a large accounting firm, Barnes and Misick, doing business with farmers in central Ohio, and received verification that many, if not most, large farmers have off-farm income from another job, another business, or another investment.)

Can a man make a decent living on a small farm? "Oh, I think he can," says Lessig. "You might want to get a little off-farm work to tide you over the rough spots. You might have to accept a so-called lower standard of living, by urban standards. But there are still many farmers who derive a good living from thirty to forty dairy cows and 100 acres. You don't hear about them because they are quite self-reliant and don't spend a lot of money for all the things farm advertisers tell them they have to have."

Mention the word agribusiness to Lessig and his smile disappears. "I don't like that word. Have you noticed how farm magazines are substituting 'agribusinessman' for 'farmer'? I don't think agribusiness really favors the farmer. It favors big business. Power is what the game is about. Agribusiness would prefer to deal with just a few farmers who are heavily indebted to it, not a bunch of small independent farmers who won't fall in line so easily. Before it's all over, we'll be just like Russia, with the only really efficient food production on small, private holdings."

Small farmers like Lessig are attracted to more natural programs of crop production, away from chemicals. "It's for two reasons, I guess," he says. "First of all, on a small farm, you can succeed only by keeping costs down. You have to think about how not to spend cash. I use less chemicals and all the manure and natural fertility I can get because I think it means keeping more of the money I do make. But also, I'm convinced that the technology of "bigness" today is a step backwards. I believe that the old ways were far better for everyone, and I've done it both ways."

Some small farmers have turned completely away from chemical technology for similar reasons. "I've gone to total organic farming," says young Johnny Adams in Bucyrus, Kansas. "I can't get the yields organically yet that we were getting with chemicals, but I just couldn't enjoy farming chemically. If you are close to the soil, you have no trouble switching to organic methods once you see what anhydrous ammonia does to soil life."

Although they don't show up in USDA statistics, small farmers are the backbone of American agriculture. Their yields often are not as high as large farms; their output per unit of energy input is always higher.

Other young farmers with mixed feelings about chemical agriculture, still hang on the fence, wondering if they could succeed with natural methods. Mention organic farming to brothers Dan and Leonard Hostettler of Mineral Springs, Arkansas, and their eyes sparkle with interest. "We've been engaged in custom farming with our father, and now are starting our own farm. We'd like to do it right organically and avoid the waste and hectic pace we often observed in large-scale farming." The Hostettlers tried an acre of sweet corn for retail sale last year as an experiment. "We sold all the corn in one day!" says Leonard. "So we think a roadside stand or pick-your-own business might make a small, organic farm profitable. We know a grower in Indiana who does all right with blueberries."

Missouri farmer Gene Poirot has been practicing and preaching sound ecological farm management for most of his seventy-six years and is still going strong. Thousands of visitors stream through his farm in Golden City, Missouri—congressmen, educators, students, sportsmen, farmers—to see why Poirot has won a whole fistful of soil and water conservation awards. "I was a life-long friend and coworker of the late Dr. William Albrecht, agronomist at the University of Missouri. We proved on my farm that his theories, though snubbed and ignored by agribusiness, are correct. Albrecht was very critical of the way chemical fertilizers were and are being used—not against them, but against the excessive way they are applied to stimulate plant growth for quick profit, rather than to restore mineral imbalances in the soil. Albrecht also maintained that the use of pesticides was an act of desperation in a dying agriculture, that insects and disease were the symptoms of a failing crop, not the cause of it. That kind of talk is not relished by the powers that be. Albrecht's methods would put too many agribusinesses out of business if adopted universally."

When Poirot started farming in 1922, he faced a monumental task. "The soil was so poor my first corn crop barely made the seed back. There was nothing there to build organic matter. I added minerals in any form I could get. I developed a process for making artificial manure out of straw and water. I built ponds for irrigation. Nature has been so mistreated here she had to be helped before she could help herself. Now, with fertility restored and organic matter accumulating rather than depleting, we use less fertilizer, not more. And we have controlled insects biologically for forty years."

In place of a worn-out farm, Poirot's acres today teem with life. Crops yield seven times what they did in the twenties. Herds of healthy cattle roam the pastures—"disease-free for thirty-two years, ever since we got the soil minerals balanced properly." The water running through his farm in Coon Creek is clear and full of fish. "In 1922 we seined two miles of the creek and netted only sixty-four pounds of fish. Today you can easily catch more than that in one afternoon with hook and line in one hole."

Wildlife workers count nearly 30,000 ducks wintering on his ponds and feeding on corn left after harvest. In 1922 there were no ducks. There were no geese then either, but now 8,000 winter on the farm every year. "According to local records, there were no deer here for 100 years," says Poirot. "But in 1970 a doe joined our cattle herd and gave birth to a fawn.

"These benefits of wildlife, soil, and water conservation—and the ultimate benefit of a healthier humanity—are by-products of good restoration farming," he continues. "They come as rewards, you might say, but my first reason for farming this way is to save money. Hedge fences are cheaper than steel posts; ponds are cheaper than wells; biological insect control cheaper than insecticides; sweet clover cheaper than chemical nitrogen; feeding hay in the pasture cheaper than hauling it to the barn and back, and hauling the manure out. And each practice in restoration farming results in many other benefits. The ponds produce great amounts of fish protein. The water in the pond becomes rich in fertilizer value from the fish manure. The hedge fences shelter birds which eat insects, and insects which eat other insects. The more you work with nature, the more nature works for you."

The future? Small farmers and organic farmers remain optimistic in the face of predictions of their demise. They point to the precarious financial position many large commercial farms are in. "To expand, many farmers have borrowed more money than they can possibly pay off in a lifetime, and they freely admit it. They bank on inflation to keep them alive. A depression would wipe out many of them overnight," points out Ralph Engelken, who's farmed 700 Iowa acres for eighteen years. "But more to the point, when you are paying $100 to $150 per acre just in interest on a large farm, you can't afford any setback. You have to keep on using chemicals whether you need them or not, just for insurance. You certainly can't afford to risk a changeover to healthy soil which might mean a couple of years of lower yields. But if you don't change over—well, I just think that if you keep going for more and more chemicals, eventually you will go broke."

"I'm kind of old-fashioned," says Paul Lewis, a small organic farmer in Louisburg, Kansas. "I think we're going to see more smaller farms in the future because of labor. Despite all the fancy big machinery, the big farms still have to hire labor, and that's a real problem in agriculture. Good labor is hard to find and cheap labor can cost you more money than good labor. Remember how everyone got excited fifteen years ago about cow pools, giant dairies where thousands of cows would be milked? Most of those experiments didn't work because hired help that will take care of cows profitably is extremely hard to find. Even when big farmers can find someone to do the work for them, it runs their cost of production up. If you think food is expensive now, did you ever think what would happen if we got union labor in all of farming? The small farmer doing his own work is much more efficient."

Other farmers point to the apparent limits of chemical efficiency as well. "There's a point beyond which pesticides won't pay," says E. J. Dietrick, president of Rincon-Vitova Insectaries in California, a firm offering biological pest control to farmers for ten dollars an acre. "Most of my work now is with big agribusiness farms like Hunt-Wesson Foods. They are learning you can't bludgeon nature into line with 100 percent chemical control. It costs too much money."

What small farmers emphasize most of all is the happiness they derive from their way of life. Roger Williams, who makes a good living on an eighty-acre Wisconsin farm, says in *Countryside*, the up-and-coming magazine that addresses itself specifically to the small commercial farmer: "Anybody can do whatever he really wants to do, including making a living on a small farm. A lot of people have the impression that a small farmer is a poor, hard-up fellow using far-out methods and living in rustic, primitive surroundings. I feel they're wrong." But, points out Williams, money-making is not the small farmer's main interest in life. He likes to spend a lot of time with his family. He enjoys little things like the nesting birds, the geese passing overhead. "There's something about the first flock of geese in the spring. Maybe other farmers wouldn't care, but if I were in a tractor cab instead of working with my horses, I'd miss the geese.

"With a bigger farm, you're dependent upon many people and many others depend on you. I like the idea of being free. I think I'm as free as I can be."

Gene Logsdon

Chapter 1

Today's Organic Farmers

Getting Started in Organic Farming

Editor's Note: Jim Foote is in his third year of organic farming. In October of 1974, after serving for four years as Farm Services Director for Organic Gardening and Farming®magazine, he and his wife

27

Barbara and their two children moved from Pennsyl-
vania to Kentucky to begin farming in partnership
with Jim's father.

Their operation consists of 255 acres of land, 30 of
which they own and 225 acres that is farmed in
partnership.

Looking back to that brisk October morning almost
three years ago, riding along on the 530 and prepar-
ing the soil for what was to be the first planting
of winter wheat on Basin Spring Farm in many a
decade, I guess I would describe my feeling then as
one of guarded optimism. I had plenty of doubts and
unanswered questions, but then I also had certain
advantages in starting out farming because of the
partnership with my father. He had the land, which
was in a fairly high state of fertility, a basic comple-
ment of machinery, and a small herd of beef cattle.
Our immediate needs consisted of a place to live and
capital to get started.

It took us about two months to get settled, move in
a house trailer, and hook up all of the utilities.
Reluctantly, we settled on a mobile home as the
simplest, least expensive solution to finding a place
to live. We found bank financing to be the cheapest,
most available source for all our credit needs.

Once settled, our basic aim was to intensify and
diversify the production on the farm using organic
farming methods. Additionally, we hoped to create a
model of diversified, long-range family farming and
tenure transfer for this area and type of farming. We
analyzed in great detail what we had to start with,
where we planned to be at the end of the first year,
and what we needed to get there. That December, I
spent a lot of time at my desk, calculating various

production possibilities and breaking down management, capital, and labor requirements on a month-by-month basis. Because timeliness is so crucial to getting things done on the farm, I didn't want a month to go by with missed opportunities.

In doing my planning, I was especially determined to avoid several common mistakes I'd seen other farmers make. These included 1. making the wrong choice of enterprises; 2. letting enterprises "get away" from them; 3. lack of diversification; 4. marketing strategies not well thought out; and 5. having too small a volume to make the effort worthwhile.

Much of my determination to pursue a mixed farm economy was based on my observation and admiration of my Mennonite and Pennsylvania German neighbors over the preceding years. By avoiding the monoculture approach, I'm not likely to make a killing on any one of my crops, but I'll always have something to sell in good years and bad. Then too, keeping a happy balance between livestock and crops and maintaining careful rotations are basic elements of my organic fertility program.

The plan that evolved is based on 60-acre sections. We had about 180 tillable acres, and we decided on a rotation of 60 acres of row crop, followed by 60 acres of small grain and 60 acres of meadow. This would begin to intensify the production and also head us toward the diversification that I think an organic farm must have. The number sixty also carried into our livestock plans, as we hoped to produce about sixty head of hogs twice a year and sixty feeder calves from the feed we grew.

Along with these first-year objectives, I worked up a three-year plan designed to bring 150 acres into three-year rotation. I allocated my fields as follows:

TABLE I-I
PLANNED THREE-YEAR ROTATION

FIELD	ACREAGE	1975	1976	1977
Hill	38	8 Wheat 10 Beans 20 Corn	8 Hay & Straw 30 Wheat	Hay
Racetrack	21	Beans	Wheat	Beans
Creek	35	Hay	20 Corn 15 Beans	Wheat
Graveyard	36	Hay & Pasture	Hay & Pasture	20 Corn 16 Beans
Saratoga Bottom	8.5	Beans	Wheat	Corn
Front of House	8.5	Hay	Corn	Wheat
Saratoga	4	Hay	Corn	Wheat
Total	150			

By figuring my acreage this closely, I was able to predict average prices and yields and come up with fairly reliable estimates of projected income. While I'm not worried about amassing any fortune these first few years, I do mean for my efforts to generate enough income to support my family comfortably while buying the farm, and to enable me to farm the land in a way that will enhance, not diminish, its basic productivity.

In order to get started on this plan, several things were needed. We had only one tractor—a John Deere 530—that was modern enough for a variety of work and I knew that we needed another tractor to meet crucial schedules and also to avoid being out of a tractor in the event of breakdowns. One more tractor needs one more plow, and our list of machinery needs began to grow:

1) Second tractor
2) Second plow
3) Manure loader
4) Corn planter
5) Combine
6) Hay rack and baler

The John Deere 1530 tractor and manure loader were purchased new because of their importance. We managed to locate a good second-hand plow to match the tractor, and also bought used the combine, rake, and baler. We finally had to buy a new corn planter after watching all of the late winter and spring for a good used one.

It's been said that the first money a farmer should spend is on land improvement. That's where we had an advantage because a lot of that money had already been spent. Our cropland had been in a thick fescue sod for the last several years. Liberal amounts of limestone had been applied to most of the land, and some of it had received phosphate—a nutrient that is short in this area.

Our only fertilizer application the first year was four tons of limestone per acre on some of the poorer soybean ground and the inoculant for the soybeans—if you want to call that fertilizer.

We had a terrible season—dry weather hit us in late July and lasted until the first of September. The corn was hit just when it needed water the most. Yet we estimate our corn made over eighty bushels on the better ground and the soybeans averaged thirty-seven and a half bushels per acre over the scales. Both of these yields are about average or slightly above average for the chemical farmers farming equivalent land in this area. The soybean yield was achieved in spite of one six-acre field that made about twenty bushels per acre after a neighbor's cows got into it.

I think the high porosity of the sod fields helped us catch the early moisture to carry us through the dry weather. This, along with the high organic content of the soil and the fact that the fields were worked on the contour, made water less of a limiting factor for us than for some of our neighbors in this dry year.

Jim Foote cultivated his corn rather than apply an herbicide. All the corn was cultivated once, some of it twice. His yields were not hurt at all by weed competition, and area farmers were impressed with the increased water-holding capacity of the cultivated fields during the dry year.

I would have liked to have applied some fertilizer, especially to the corn ground, but my manure supply wasn't established prior to this first crop. We hadn't fed any cattle or hogs. And the prices of all commercial fertilizers were so unreasonable in the spring of '75 that I decided not to apply any phosphate. Fertilizer prices have since dropped to more realistic levels, I've located a supply of chicken manure, and our own manure supply is accumulating for the coming crop year. The way things worked out with the dry weather, I think we did well by not shooting for high yields that first year.

It might be argued that the lack of manure and rock phosphate hurt us some. We'll be able to tell more by the response we get this year, assuming conditions are about the same.

Our weed control effort was aimed at getting all of the weeds we could without the use of chemicals. Notice I didn't say *all* of the weeds like some of the experts stress. But I'm more than satisfied with our weed control this year compared to the best chemical farmers around. Corn and soybean ground was plowed in the early spring with a moldboard plow and planted as early as possible. I considered delaying planting until later in order to destroy more weeds with the disc harrow, but decided to plant early instead. The soybeans were cultivated once with a rotary hoe and twice with a spring-shovel cultivator. Part of the corn was rotary-hoed, about half was not. All of the corn was cultivated once, most of it twice. I couldn't tell much difference as long as it got one good *early* cultivation. Weed control in the soybeans was excellent—as good as I'd hope to see. Weed control in the corn ranged from fair to good. We had no crop that suffered from weeds. I found one trip over with the rotary hoe and two trips with the cultivator gave me excellent control of weeds provided the first cultivation was timely.

I stress this point because I've had several farmers tell me that I couldn't handle the weeds without chemicals. A bad job of cultivation compared to a good job of chemical application is unimpressive to me. But show me a good job of cultivation and I'll take it every time—particularly on slightly rolling ground. Contour cultivation does a great deal to improve water absorption and reduce soil loss.

Weeds were controlled in the pastures by clipping. Part of the pasture was clipped with a rotary and part with a sickle-bar mower. I lean toward the rotary mower because of the way it shatters the residue, leaves a taller stubble for cattle to graze, and requires less maintenance. However, it does consume more fuel.

A neighbor asked me if I thought I saved money by cultivating rather than spraying my crops for weeds. I told him that was not my intention. However, my cost per acre for weed control in row crops was around five dollars, figuring my labor at three dollars per hour and fuel at fifty cents a gallon. I doubt that chemical weed control would have been much cheaper, if at all. There would, however, have been a saving in time spent with chemicals—a factor that should never be underestimated.

There are other, nonfinancial benefits from cultivation that should be considered. I planted my crops on the contour, and cultivation with the shovel cultivator created little "terraces" that helped slow runoff, reduce erosion, and increase water absorption. Also, cultivating my crops gave me the best possible supervision of their progress, at a time when I was unlikely to walk over them as thoroughly as I like to. Since the sod grasses were not killed by a blanket application of herbicide, those that persisted in the row atop my little terraces helped hold the soil during the early summer rainstorms. These benefits are not to be ignored.

The toughest weed we are faced with in this part of Kentucky is Johnsongrass. We haven't really confronted this yet, and the problem gets worse with intensive row-cropping. Johnsongrass makes excellent forage, and doesn't appear to be a problem in pasture land, but it can destroy a corn field. We've also "imported" a few weeds (namely, giant foxtail and cocklebur) not native to this area in what was supposed to be clean, certified soybean and clover seed. Live and learn.

In the early spring when the weather wasn't suitable for working the land, we started working on fences. We improved the cattle fences, made one field hog-tight in order to pasture brood sows, and built a small lot with access to the barn as a place for finishing pigs. We started out by purchasing eight sows and a boar. We wanted the pigs to start arriving about the time the corn crop was available, and this worked out pretty well, with the first litter coming in late August.

The eight sows raised sixty-three pigs to weaning—one sow failed to breed. So we averaged nine pigs each from the sows that farrowed. This sow herd will either be increased or disposed of in the future. It might be that we can do better by purchasing feeder pigs if disease isn't too much of a problem.

Cattle are still the main enterprise on the farm. Much of this country is eminently suited to the raising of grazing animals. There is abundant water from springs, ample shade from scattered trees, and much of the land is so rolling that it should not be cultivated often—some of it not at all. Grazing animals represent the best use of this type farmland.

We are running about 50 head of brood cows at present and hope to increase this by about 10 next spring. This means that at any given time we have from 95 to 125 head of cattle to feed, including calves, yearlings, bulls, and replacement heifers.

These cattle subsist entirely on pasture and hay.

Salt is our only purchased item for the brood cow diet. For this reason, a great deal of effort goes into maintaining the supply and quality of pasture and hay.

The chief grass in our pasture program is Kentucky 31 fescue. This hardy grass provides abundant forage from early spring to late fall. It does go through a summer dormant period, but at that time the perennial legumes help fill in.

The trick with fescue is to keep it from completely dominating the pasture and smothering out the legumes that provide our pasture with nitrogen. Right now we need to renovate several acres of pasture and reintroduce legumes. Red and white clover are the ones we depend on, although I will also be trying some alfalfa and birdsfoot trefoil this year. Clover is interplanted with spring-sown wheat in my rotation, and I've also had success with an alfalfa-orchardgrass-oats combination. With this companion grouping, if the oats are not worth combining in a dry summer such as 1975, the cattle can be turned in for some quality forage.

Clover, agricultural limestone, and phosphate are the key ingredients in our pasture fertility program. The manure is returned to the soil in a pasture program but it never seems to be as beneficial as when it's spread on cropland. The nitrogen is largely lost, and the scattered droppings soil the pasture so the cattle won't eat it as well.

Our primary concern this first year has been to work on the production end of our operation. We've yet to market anything as organically grown, and we plan to work on this aspect of our business this winter. Right now we have about seventy hogs on feed and twenty-two steers that will finish out in early spring. These animals will do well on any market, but I'm interested in selling them to people who want to know how they were produced. We raised each animal from breeding to feeding and they're consuming grain produced here on this farm—without chemicals. I'd like to see more of them going to people who care about these qualities in their food.

I really think a person ought to see his meat animal on the hoof if he's buying for the home freezer. It represents a sizable investment to fill a freezer with beef, and there's a great difference in quality. I'd like to interest my customers in looking at their animals—especially the first time they buy.

This goal of marketing our products as organic still looms as one of our greatest challenges. Trying to develop a market for organic beef or crops in Breckenridge County, Kentucky, is a far different proposition than when you're farming close to the East or West Coast population centers. It would be fatal to make one's plans on the basis of any anticipated price or demand bonus from being organic.

Not surprisingly, the indifference to organics carries over to the neighboring farmers and the state ag extension people. It's not that they've been hostile—it's actually more of a detached amusement and wait-and-see attitude that is characteristically accorded any new venture in these parts. The older farmers have seen many ambitious, bright-eyed young farmers come and go. Their skepticism is well-founded. Some of our brightest moments this past year have come when one of these older farmers has allowed that our corn or beans "weren't so bad, considerin'."

As for labor requirements, the cropping plan I've described seems to work out about right for my father and me. We do some work swapping occasionally with our neighbor, Shirley Wilson, and hire local high school boys for haying and other peak periods.

From the start, Barbara has kept meticulous daily

records on our expenditures and income. Detailed records are an absolutely critical tool for planning the farm's future. One of our goals for next year is to find and cultivate a permanent relationship with a "personal" banker—someone at one of the local banks with whom we could review these cash flow figures and discuss various approaches to our short and long range goals.

The real challenge in formulating all these plans is to find a happy medium between being too rigid and too flexible. For instance, I'd like to stick pretty closely to my sixty-acre rotations, but if my studying and intuition tell me soybeans are going to stay at six dollars a bushel next fall, I'd be foolish not to plant an extra five acres. Hog prices, on the other hand, have dropped depressingly low, but I don't plan to withdraw from the hog business altogether because of it.

These decisions—and they have to be made daily—are at once the agony and the satisfaction of farming. Each day provides new opportunities for exploring just how much one can do to improve the farm and its productive potential.

Joe Hamada: Truck Farmer Extraordinary

One close look at Joe Hamada's soil is all it takes to know why he's so successful at growing cash crops all year-round. The fifteen-acre plot of sandy loam and adobe shows clear evidence of the care Hamada has lavished on his land. In turn, it brought him a total of $50,000 in premium-quality produce in 1975, and 1976 promises to top that.

Set about two miles inland from the Pacific and

Joe Hamada farms his soil 12 months a year in southern California. He has nicknamed his farm God's Green Acres, and following his 27 years of effort, the name fits it well.

twenty-five miles north of San Diego, the hilly farm site near Leucadia keeps up a steady return of vegetables, one bumper crop after another. And just as steady is the return of organic matter—poultry wastes, plant wastes, fish meal—which keep it productive. That's Joe Hamada's secret. Despite around-the-calendar cropping, his soil stays alive, rich in humus, and ready to keep the plant-producing cycle going.

Hamada's prime crop is leeks. Between five and six acres of the big onion-family members are grown continuously. "It's a crop you can leave in the ground if the price or demand isn't right," explains Joe. Other root vegetables are also favorites: beets, turnips, carrots, along with sizable plantings of lettuce, cabbage, beans, kohlrabi, Swiss chard, tomatoes, green onions, yellow and zucchini squash.

Most of the truck farm's steady harvest goes to two wholesale organic food distributors in Los Angeles. Some reaches the local community by way of sales at a Leucadia "country market," while some gets to people who come right out to the farm to buy it.

Like his father before him, Joe Hamada has been truck farming all his life. As a youngster, he worked in his dad's fields in Long Beach, where his parents had come from Japan. In 1932, as a teenager just out of high school, he took over the family plots in Orange County, and since 1950, he's been one of the foremost growers in the organic food movement centered in that area.

"I was a chemical farmer at first," confesses Hamada, "but when I saw what was happening to the soil, when I saw the detrimental effect of poisons, the increase in insects instead of controlling them—I became an organic farmer as my father had always been." In the 1930s, when farmers started using chemical fertilizers, "they got by with 200 pounds per acre," Joe told me. "Soon they had to double that amount, then double it again. Those fertilizers brought in many other problems—new diseases, attraction of insects. The pests that came then called for insecticides, and these have had to get stronger and stronger as more resistant strains come along. The powerful insecticides kill natural organisms in the soil."

In contrast, Hamada has worked to encourage the balance of nature he saw being upset. A deeply religious man, he deplores the destruction by farmers of "the balance that God has established. I have found out that the less I do in the fields, the better it seems to be. The birds avoid insecticide—they know where it is. Birds swarm to my fields and eat the bugs. Many farmers have problems with aphids. In the natural state, ladybugs eat the aphids, but where insecticides are used there are no ladybugs." He also uses wild marigolds and has tobacco surrounding part of his acreage, which he lets stand as a further natural barrier to pests, and said that gopher snakes help out.

What Hamada does do in his fields is replenish generously for succession planting. He relies mainly on chicken manure from nearby poultry operations, which he says provides a higher NPK than other wastes. Along with it, he uses a modest amount of fish or blood meal and ground rock phosphate for crops that draw most on nitrogen and phosphorus. On a plot about to be planted, he usually spreads and turns under about three truckloads of chicken manure—approximately thirty-six cubic yards to the acre. There's also an overhead irrigation system for 95 percent of the farm, with drip irrigation being tried out on the rest.

The rotation that Hamada follows in producing year-round yields is a careful one. He alternates carrots, beets, leeks, and other root or underground crops with cabbage, squash, broccoli, tomatoes, and different vegetables that grow their edible parts above-ground. The combination of waste recycling and rotated fields brings him three to four crops a year everywhere on his land, depending on the maturity time of the succession. Five workers hand-cultivate weeds the year round—with Joe, his wife Toshiko, and some of their eleven children helping out.

"Of course the result we're trying for is healthful, natural vegetables," said Hamada. "People tell us our vegetables taste like they used to in Grandma's time." It's no wonder they do—the way he grows them. The land has been good to Joe Hamada, an energetic, sixty-three-year-old truck farmer, because he's been good to the land.

M. C. Goldman

For Eliot Coleman, the key to a successful Maine truck-farming operation is "don't spend more than you earn."

A Small Farmer's Guidelines to Success

After ten years of experience in making a living from a few acres of land along the Maine coast, the key to successful small farming to me is "Don't spend more than you earn." Early in our experience,

I found out I had to ignore economists. Most economists would have you worry if you were getting a fair return from your labor, but the real critical measure is not whether your labor earns little, but that you pay out a little less.

Here are ten rules for success in small farming and large-scale market gardening:

1) Own the farm. No mortgage payments would be ideal, so it's essential that you buy only what you can afford. In New England, don't buy any more than forty acres. A forty-acre farm will provide you with fifteen acres cleared and twenty-five in woods. Even at a slow rate of forest regeneration (one-half cord per year per acre) you should have a self-sustaining supply of firewood and fence posts, etc. Fifteen acres of cleared land intelligently farmed by one family will yield a more-than-adequate income.

2) Build once and build well—barns, outbuildings, etc. Know what you want and can afford. Temporizing wastes time and money. Wait until you are organized both physically and financially before beginning a new operation.

3) Avoid monthly bills. If we run out of money in March and have to wait several months before the lettuce crop will start bringing in dollars, we can eat out of the root cellar. Electric bills, water bills, telephone bills are eliminated in our way of living, and it's rather fun being as independent of cash economy as we are for those few months.

4) Be as *diversified* as possible. This year, our strawberries froze out so we were $300 less, but our raspberries became the best crop and made up for the loss. Diversify with compatibility—using good business techniques. We have built up a reputation with our customers that we have everything they ever wanted in a vegetable garden. We start some crops extra early—cucumbers, peppers, so we can sell them by the Fourth of July. We grow forty different

varieties of vegetables—some extremely difficult to find for most people, like leeks and salsify. We're always trying to escape weather troubles by planting some crops that need hot weather, and some, like cabbage, which need cool weather.

5) Quality production is vital. I am not forced to sell inferior produce to my customers. The main thing is to be able to sell what you grow—*with pride*. Run your farm like a business. If it isn't a successful business you won't make it. Produce the highest quality possible and it will command a good price. Search out new markets. Triple-crop or at least double-crop for efficient land use. If you run a market garden acquire the reputation that if anyone will have it, you will have it—and then have it.

6) If you can grow it or make it, don't buy it. TV is designed to sell the viewer products, so the less you watch it, the better. That same advice goes for organic fertilizers like rock phosphate. If you can make use of the phosphorus in your soil instead of buying rock phosphate, so much the better. For animal feed, instead of buying it in the feed store, give your livestock crops you can grow, such as small grains, carrots, mangels, parsnips, pumpkins, squash—the seeds as well. The corollary to this rule is not to sell only raw materials. For example, process milk into yogurt or cheese.

7) Don't make artificial rules. I am always skeptical when I hear someone say "I'm not going to use any motorized tools." I am firm in my convictions about not doing anything to destroy the world, but I also believe in not making things tougher. I dig some muck from my pond because the organic matter improves my soil faster and helps get the clover established in the pastures. I am not digging enough muck to change the ecology of the pond, although some people have criticized me for using any of it at all. Also don't become "converted" to a system.

There are many successful ways to farm and many techniques. Don't practice one method blindly (i.e. mulching, deep-digging, etc.) to the exclusion of common sense. Learn from all methods and apply the technique to the situation, the crop, the soil, the weather, the farm, or whatever. Be flexible. Seeds are determined to grow and do not require any magical "method" to help them.

8) Study, read, analyze, and keep notes. Your own personal experiences can be most valuable in helping you decide crop rotations, which vegetables to grow in the coming year, etc. See what materials your state library offers. Get pamphlets from the extension service, the government, foreign countries. Keep extensive notes on what grew well and where. How was the soil prepared? Try new varieties in small tests every year. Be continually observant when you walk about the farm. The old saying goes, "The best manure for any farm is the footsteps of the farmer."

9) Plan ahead and plan well. Just because you are a small farmer doesn't mean that you can be careless. When you have to move materials or heavy objects move them once to where they are going to remain rather than to a temporary spot from where they will have to be moved again with a waste of time and effort. Make an extensive long-range plan of the work that needs to be done and then plan it for the most efficient use of time and resources.

10) *Believe* you are going to make it. Faith and confidence are extremely important. We have three and a half acres in intensive vegetable production, and just ten years ago, I planted our first garden. There is so much more available knowledge, tools, and technology today than 100 years ago when farms were abandoned. The 5-cow dairy farmer does not have to be duped into the so-called "economics of scale." He can do better than the farmer who has 50

cows, and then is told he should have a 100-cow dairy herd. The 5-cow dairy farmer should sell butter, cheese, and yogurt.

All of us have agriculture in our blood. We may have to go back five generations, but it's there and that knowledge also should make you extra confident.

Eliot Coleman

Steve Beck (center) has been farming 2,000 acres of dryland wheat in California for four years.

Organic Wheat Grower in California

When you say dryland farming to Steve Beck, you can be sure he knows just how dry it can be. From early March until the waning days of August 1976, not a solitary raindrop fell on his thirsty 1,000-acre spread of wheat. By that time, it did little more than settle the dust. Yet, despite California's worst drought since 1852, he's harvested a modest crop—some 250 tons of exceptionally high-quality grain—enough to carry him into a fourth year of organic production.

By definition, dry farming constitutes what is practiced in "regions having little rainfall and no irrigation, by using methods which conserve soil moisture and by raising crops which survive drought." That description fits Steve Beck's situation and his stubborn effort exactly. Rainfall in the Carrisa Plains of Santa Marguerita usually averages a meager eight inches annually. This year it barely reached a stingy five-inch total. As for irrigating, "We couldn't do that even if we wanted to," he explains. "There's no water here anyway."

What *is* there is a stretch of parched clay soil between the Templora Mountains to the east and the Santa Lucia coastal range on the west. At a 2,000-foot elevation, the land bakes in summer heat that hits 115 to 120 degrees, then plunges to zero in the winter, with snow capping the surrounding peaks.

"Most people don't believe wheat grows here," Beck says, and while per-acre yields are considerably under the twenty-five to thirty bushel returns (1,500–1,800 pounds) common to grain-belt states, growers in California's hot plains bring in harvests of good quality wheat ranging upwards of half that amount. What's more, its nutritional value is high, a plus-factor in producing hard red winter wheat for bread and other flours. "Lots of stress such as plants face with the heat and wind chill here produces higher protein content," says Steve. Lab tests on his hard red crop this season show protein at 14 to 16 percent, while some of Beck's Durum wheat analyzed as high as 19 percent protein.

"This is the kind of year you can tell *how* the grower has taken care of his land—if he has a crop!" exclaimed Steve Beck as we walked through some of the short INIA-variety winter wheat on a 117-degree July afternoon. Ordinarily, this strain grows about three feet high (most wheat reaches another foot), but this year's drought has kept it to little more than half that tall. Just the same, the grain heads were pretty full.

Nearby growers haven't done as well in the drought, Steve noted; some have gone in the red on 2, 4-D weed-killer and other chemicals with the low-yield season. It's the way Beck treats his soil and the way he approaches farming itself that makes the difference. Three years ago, the young Sonoma State College graduate and his wife Debbie decided it could be done. "I wanted to grow wheat organically because I don't like to see land or crops sprayed," Steve said. "And I don't want to be just a cog in the huge commercial-agriculture machine when I market what I grow. My neighbor farms the same land, but he doesn't farm it with his heart." Adds Debbie, "It's a lot of hard work—Steve puts in an eleven-hour-plus day."

The system Beck has worked out is called a summer fallow. Actually, he's farming on 2,000 acres, alternating tracts each year and allowing the soil to rebuild in the off, or fallow, year. He tills right after harvest, and again four times during the fallow months. Good tillage, both then and before planting, he explains, gets around the Russian thistle weed problem—without poisons. It costs more to till that much, and yields may be somewhat smaller, but there's no bill for chemicals, no residues in his soil or in his wheat.

Two months after harvest, Beck has sheep and cattle graze the fallow half of his land. About 1,500 head of livestock from nearby ranches, to whom he leases grazing privileges, help put the resting fields into condition for the next growing year. Combined with the weed and plant residue tilling, it's all the fertilizing his soil needs—and it helps hold moisture through even tough drought years like the one just experienced.

Once harvested, Beck's wheat crop is stored in 100-pound sacks, bearing his own Carrizo Grain Co. label, then in twelve-foot elevated bins one and one half feet off the ground. The almost totally dry environment means little or no trouble with field insects, and less critical problems with any in storage. Diatomaceous earth (Perma-Guard) helps avert any weevil damage in the bins.

Besides the hard red winter wheat, Beck grows some soft Durum, which needs more moisture, suitable for noodles and other pasta products. He also plants a field of barley, some of which is sold to the Vitova Insectaries in Riverside to be used for rearing parasitic insects. Everett J. Dietrick of the biological control firm says he's glad to have a supply of organically grown barley, since pesticide residues had previously cut their breeding rates.

The bulk of Beck's wheat crop is sold to the Erewhon Trading Co. of Los Angeles and Boston. Delivered price for the hundred-pound bags has been about $10.55, he reports. Steve has bought his land from his father, and with a total of 10,000 acres in the family, he points out that there's a potential capacity for five times as much organic wheat as he's now producing, providing there'd be a market for it. He's ready and well able to grow as much wholesome grain as California's droughty soil can yield.

In an age when the quality of food, and of the way it is produced, has become a vital issue, Steven Beck stands tall—a dryland farmer who grows good wheat with his head and hands, as well as with his heart.

Doing Fine in Kansas

Bennie Unruh of Aulne, Kansas, is an organic farmer who has developed health food market outlets for all his grain and beef and bypasses the traditional marketing system entirely.

In addition to producing grain and cattle on a farm that has been in his family since 1872, he has been a registered miller for several years and grinds and packages a wide range of products distributed in Kansas and elsewhere.

He operates this year-round business from a new building at his home in Aulne, a small town about forty miles north of Wichita. It brings a lot of people to town because one of his specialties is milling flour and cereals to order.

Anyone who has seen Unruh's farm, which is a few miles northwest of Aulne, can see it is ideally situated for organic agriculture. It is on a hill, where the water drains off in four different directions, ruling out the possibility of chemical pollution from neighboring fields.

The turning point in his operation came during a visit to a health food store in Salina, where he heard the owner tell another customer he was looking for an organic farmer who could provide a reliable supply of whole grain flour.

He had been milling grain for family use since 1937, putting it through a coffee grinder several times. Neighbors and friends who tasted bread made from this whole grain flour began asking for it and he ended up buying his first stone mill. But it wasn't until that day in Salina that he considered turning it into a commercial enterprise.

In addition to growing up to 150 acres of hard red winter wheat a year, Unruh also produces corn and rye for his milling operation. He gets good yields of up to 100 bushels an acre for corn, which is used for both feed and corn meal, and up to 50 bushels for wheat. He credits the regular application of organic fertilizers for the high productivity of his soil.

He recalls the first time he contacted the company and reported that the family farm appeared to be run down. It had been cropped for nearly a century without anything except manure being applied to the fields. A Hy-Brid Sales Co. representative recommended some Calphos, which got results on all but one field. A soil test showed it had a serious potassium deficiency.

Although Hy-Brid Sales Co. recommended 500 pounds per acre in one area and 1,000 pounds in another on that field, Unruh spread a ton on a third area to see what would happen. The response from this heavy application of granite dust was spectacular and he was turned into a granite-dust enthusiast and one of the company's best customers.

Like most organic farmers, he also uses legumes to put nitrogen back in the soil. He uses a lot of sweet clover, seeding it with wheat in the fall and letting it grow to full maturity before working it into the soil.

His operation also includes a herd of twenty stock cows and about the same number of steers and heifers being fed out on organically grown grain. "They're all spoken for," he said, referring to standing orders from customers for cattle from his feedlot.

He has about fifteen acres of alfalfa, which is important in his livestock operation. He noted that alfalfa needs minerals and phosphates and said he normally uses 150 pounds of Calphos and 150 pounds of granite dust per acre at seeding time.

Unruh said he plants seed varieties especially suited to milling and keeps back enough for his own use so he doesn't have to use chemically treated seed. He emphasized that he has never had any mosaic in his wheat, a problem for chemical farmers in his area.

Roger Blobaum

Five Wisconsin Farms

Editor's Note: Stein Goering interviewed Wisconsin farmers to test the practicality of organic farming. His report for the Ocooch Mountain News *is reprinted here.*

Although their farming practices vary, the organic farmers I interviewed all rely on the basic tenet that soil fertility can best be built by increasing the soil's content of organic matter—humus and soil organisms.

Hickory Hill farm, on Highway 23 south of Loganville in Sauk County, is 250 acres operated by Harold and Carla Kruse and their children. They raise corn, oats, and hay on their 150 tillable acres and milk twenty-five to thirty Holsteins. Composted manure from the cows is hauled onto their corn fields; they also apply a mixture of rock fertilizers (Calphos, which provides phosphate in a collodial form, and Hybro-tite, a source of natural potash) and trace elements and additional nitrogen in the form of dried kelp. The Kruses buy the separate ingredients and mix them in a cement mixer. This extra work saves them money compared to the cost of the commercial blends of similar materials. An application of 500—1,000 pounds per acre will provide nutrients for several years since most natural fertilizers do not leach away.

A seven-foot rototiller and a Graham chisel plow,

pulled behind their Fordson diesel tractor also help to increase the humus content of their soil. The tiller mixes plant matter with the top layers of soil while the chisel aerates the lower levels—rather than just burying crop residues with a moldboard plow.

Although its mainstay is the dairy herd, Hickory Hill is a diversified operation which begins to approach the ideal of a self-sufficient homestead. The Kruses have turned a rocky sheep pasture into a highly productive, three-acre truck garden. They market surplus produce at their roadside stand, and also through the Farmer's Markets in Madison.

They tapped 150 maple trees and boiled down their own syrup; they also grow cane sorghum and hope to have their own sorghum mill set up by this summer. Their greenhouse is full of plants—many beautiful and exotic cacti as well as garden seedlings. They keep a flock of laying hens and there is a demand for their organic beef.

The whole family is involved with the operation of Hickory Hill. The oldest son, John, is into mechanics. He helps keep the farm machinery running and has plans to restore an old threshing machine. He plans to grow five acres of spring wheat this year.

Part of their house is given over to a natural foods store. They handle whole grains, nuts, dried fruits, natural vitamins, flour mills, and juicers, as well as the complete line of Shaklee soap and cosmetic products.

Another organic dairy farm in Sauk County is Vince and Lou Garvoille's place near Spring Green. Along with two of their sons, Vinny and Jim, they milk twenty-five to thirty-five Holsteins and raise crops on their 100 acres of heavy valley land.

They have very little grazing land, so they must green-chop forage for their herd during the growing season. In the winter, their cows are fed baled hay, silage, and a grain ration which includes a vegetable protein supplement and Microlite, a mixed vermiculite ore which provides supplemental minerals and trace elements. Vince says that using Microlite in the feed not only helps keep their cows healthy, but also increases the fertilizing value of the manure they produce by enriching its trace element content.

That manure is spread on ground that's to be planted to corn; they rotate one to two years of corn with a year of oats followed by three or more years of hay. The only other fertilizer they use is a gel-like substance known as Agriserum. They apply it as a coating to their corn and hay seed at the rate of one pint per bushel. The bacteria cultures and nutrients contained in Agriserum (the exact formula is a well-guarded secret) help promote the growth of the young seedlings, and also stimulate microorganisms and earthworms in the surrounding soil. These in turn convert soil minerals into available plant food.

The Garvoilles used to farm near Sun Prairie where they used conventional synthetic fertilizers and pesticides. The results of an experiment in which they used Agriserum in place of chemical fertilizers on part of their crops with no loss in yield, plus the fact that their new farm at Spring Green had been farmed without chemicals for sixty years, led them to switch completely to organic methods. They've had satisfactory results for four years.

Verlon and Edith Bentz are a couple from Juneau County who stand in direct opposition to the trend of small farmers quitting and moving to the cities. They both grew up in the Wonewoc area, but spent over twenty years in Milwaukee where Verlon worked as a bricklayer. In 1968 they bought a 160-acre farm near Elroy and took up organic farming.

They started out raising beef cattle and went into dairying a couple of years ago. They're now milking twenty-one Holsteins and Gurnseys. They also keep a flock of chickens, grow organic fruits and vegetables, and heat their house with wood, all of which helps to cut down cash expenditures.

About half of their acreage is tillable ridge-top land, and they follow a standard corn-oats-hay rotation. They use the fertilizer boxes on their grain drill to apply phosphates, potash, and trace minerals in the form of natural rock dust; they find that layering the various ingredients right in the box is enough to mix the fertilizers adequately.

Verlon has also used SuperGro on some of his corn—a commercial blend of rock fertilizers, bacteria cultures, and soil conditioners, and feed-grade urea as a nitrogen source. This is not a strictly organic product, since the urea is synthetically prepared. However, urea is found naturally in animal manure and is a less harsh source of nitrogen than other synthetics such as anhydrous ammonia or ammonium nitrate.

Last year the Bentzs tried another natural product known as Clod Buster. This material, which is found in deposits in the Southwest, was first discovered by Navaho and Apache Indians. They used it as a soil conditioner and it enabled them to grow corn in that very arid region. It contains a high percentage of concentrated humus, as well as humic acids, which stimulate the production of additional humus when applied to the soil. The result is improved soil tilth, better moisture-holding qualities, and release of locked-up nutrients. Verlon says that applying it to an old seeding of alfalfa made an almost immediate and visible difference. The plants looked healthier, had a richer color, and produced a better yield. Verlon also felt that the alfalfa grown with Clod Buster had a higher feed value.

Community Farm is 125 acres located south of Clifton in Monroe County. The farm is operated by the collective efforts of a half dozen people. They raise cash crops on their sixty tillable acres; they keep some livestock, but the animal products are mostly for their own use.

They've sold corn, hay, and grain (a mixture of oats, wheat, and barley) which they grew last year. Their corn was fertilized primarily with Radiant Blend, a mixture of rock fertilizers, composted manures, and bacteria cultures. With last season's adverse growing conditions, they feel that their 100-bushel per acre yield of high quality corn was some of the best produced in the area.

This year they plan to grow additional crops of dried beans and sunflowers. They've also started an acre of comfrey, setting out 2,300 root cuttings last fall. Comfrey is an herbal plant with many uses—as a medicinal plant, for teas, or as a high-protein forage for livestock. It requires intensive cultivation and heavy applications of organic nitrogen, but there is currently a high demand for good quality herbal products.

The folks at Community Farm also feel that the best way to build soil fertility is to increase the organic matter content. This year they plan to use sewage sludge and Clod Buster on their fields in addition to Radiant Blend. The sludge is available from the sanitation department in La Crosse, free for the hauling. They have a two-ton truck and hope to haul one load a day to their farm for two weeks.

Clem and Virginia Wall farm 275 acres south of Mt. Ida in Grant County. They've farmed organically since 1967, using only Agriserum, and have been applying Microlite for trace elements on some fields in recent years. They are dealers for these two products, as well as Clod Buster and open-pollinated seed corn. The Walls claim that their organically grown hay and open-pollinated corn has a much higher feed value than the chemically grown hybrid feed they produced before. Clem also says that his own health improved when he quit working with chemicals.

Clem uses a rotation of corn-soybeans-corn followed by oats and hay on their best ground. They fed over 100 head of beef cattle and keep enough cows to produce milk for their own use. (They went out of dairying last year when they could no longer sell milk in cans.) His wife churns her butter from the cream and grinds flour in their own mill. The Walls also use Agriserum on their home vegetable garden with excellent results.

After visiting these five Wisconsin farms, I, for one, am convinced that it can be done. The theories of the organic method can be applied successfully to larger scale farming. A farm family can survive without applying chemicals, and produce, in the long run, at least as much food from their land while doing it.

Ten Years Ahead of His Time

His vet says that Johnny Adams is ten years ahead of his time, and the young Kansas farmer doesn't mind hearing a little praise like that occasionally. When you're trying to farm 400 acres organically and then retailing the produce yourself, you need all the encouragement you can get.

"I work harder and get less yields when I farm organically, but the products I raise have better quality," says Adams. "I want to get paid for that. It just bothers me terribly to see my grain dumped in with common stuff at the elevator or my pure beef selling for the same price as the chemicalized meat."

Convinced that his organic meat and grain are better, Adams is taking his case to the consumer's pocketbook by selling retail direct from his farm. So far, customers agree with him—they're paying premium prices and coming back for more.

The Adamses are proceeding along three marketing fronts: selling beef on the hoof, selling some grain through a nationwide organic distributor, and—the project they are pinning most of their hopes on—selling their own home-ground flours and meals.

As of now, the beef business is the most fully in operation. Customers come to the farm and pick out the animal they want. Sometimes two or more families will buy a steer together. Adams delivers it by prearrangement to a local slaughterhouse. The butchering, cutting, and packaging of the meat from then on is between the customer and the slaughterhouse. "I sell at a premium ten cents per pound above market price," says Adams, "and no complaints. In fact, demand is so high I can't fill all the orders this year and I'm thinking of raising my price."

By modern standards, Adam's fattening ration is somewhat unique. "When I first went organic, I continued to try to feed beef the way we had always done," he recalls. "Bring them in off pasture at 600–700 pounds and put them on a full grain ration. In the old days, that meant that by the time an animal was approaching 1,000 pounds in weight, it was eating 18–20 pounds of grain per day plus supplement. Well, when I got the steers on a full ration of my organic, open-pollenated, field-dried corn, then socked the protein supplement to them, they'd go off feed. By trial and error, I discovered that they were getting too much protein! Now I never have to feed more than about ten pounds of corn per day, with less supplement. I don't even have to feed extra minerals. It looks to me like a bushel of my lower-yielding, open-pollenated, organic corn is worth

Johnny Adams is farming 400 acres in Kansas organically, and retailing as much of his produce as he can himself.

more than two bushels of commercial corn."

Adams has adopted other controversial views based on his experience. "We've learned that an Angus-Holstein cross makes a thriftier animal to fatten than a straight Angus. It also tastes better, much to our surprise. Other crosses with Angus, which is our basic breed, aren't so good. We don't think an Angus-Charolais cross is fit to eat."

Best time to market? "When your customers want their meat," answers Adams. "That could be any time, but in our experience it is never at Christmas. I guess everybody is eating turkey then."

To market grain not marketed through his beef, Adams has embarked on a more ambitious project—selling it as home-ground flour or meal. "Wheat's worth around $2.50 a bushel, but as flour it's worth around $12.00 a bushel. There's room there for a profit, especially for the small operator who can offer a nutritionally superior whole-grain flour," says Adams.

He first got interested in milling after his wife Marie won a bread-baking contest. "We had purchased a kitchen mill so we could grind our wheat for our own use," says Marie. "I used a recipe for 'overnight bread' which you don't have to knead as much. The bread was delicious and easy to make. We felt sure other homemakers would bake bread if they could get flour like we had. And so far our hunch has been correct."

"But," adds Johnny, "If I'd known what a hassle I was getting myself into with government regulations, I doubt I would have built the mill building. There's just an awful lot of red tape involved and the best advice I can give other prospective small millers is to get a good lawyer to help you through the maze. We don't object to regulation—we *want* a good clean, approved operation. The irony is that the regulators do not seem to know how to apply their regulations to small operations. Everything is geared to bigger and more."

His mill house is an attractive building next to the driveway, modeled after the gambrel-roofed barns of yesterday. Adams designed and built the structure himself—fifteen feet by ten feet and two stories high. He fears already that the building is too small, but perhaps a wise small step for a beginner. When operational, he figures he'll have about $3,000 in the structure. Ten storage bins upstairs, with a total capacity of 170 bushels, have slanted floors which will feed grain by gravity to the first floor mill.

"But we haven't found a mill yet that satisfies us," says Adams. "Commercial mills we've looked at are all too big for us, and the little home models aren't big enough. But we'll be rolling before the year is over."

In the meantime, Adams goes on producing grain with the handcrafted skill of an artisan. Corn is field-dried before harvest—"I don't like drying with artificial heat. Kills some of the good in the grain, I think. What you have left is about like popcorn. Even the chemical guys realize the fallacy of drying with artificial heat—they won't feed the stuff to their own cattle."

To keep bugs out of stored wheat, Adams relies on diatomaceous earth. "I use Perma-Guard brand, and it works better than any fumigant," he says.

Out in the field, Adams follows a strict organic program: fish fertilizers, rock phosphate, and granite dust, along with a soybean, corn, wheat, and clover rotation for green manure and nitrogen fixation. "We've got sweet clover coming up all over the place like a noxious weed," says Adams with a smile. "And I think that's awfully nice." In addition to its fertility value, sweet clover controls troublesome bindweed, Adams has found. He has problems, too, with cocklebur in his hayfields, though he gets some control from periodic mowing of the hay. "Cockle is supposed to be a sign of zinc deficiency and is nature's way of putting zinc back in the soil," says Adams. "I'm going to see if that's true." He experiments all the time. "Frankly I can't see much result from granite dust yet, nor did I get any results from humates alone when I tried them. But where I combined humates and granite dust, the wheat was better."

Organic farming is not the easy way to do the job and Johnny Adams knows it. "But I just can't enjoy working with chemicals that poison the soil," he says. "I never could. When I first heard about organic methods I was so happy to realize other people thought the same way I did. I think it's a matter of whether you really love the land or not."

Gene Logsdon

Growing Organic Flour

Ray Juhl is one Midwest farmer who sees production of organically grown grain and stone-milled flour as an emerging agricultural industry with

For Ray Juhl, putting in his own mill made his wheat-growing operation much more profitable.

strong demand and unusual growth potential.

He's so certain of this that he has built and equipped one of the nation's largest on-farm milling setups. It is located on the 2,500-acre farm near Middle River, Minnesota, where he and his son Douglas produce thousands of bushels of organically grown wheat, barley, oats, and buckwheat.

Main operator of the new mill is Randy Heinbaugh, Juhl's son-in-law, and initial production is set at two semiloads of flour, corn meal, and rolled oats a week. The output can be doubled with longer days and weekend work.

What has convinced Juhl, a careful businessman, that these large orders will materialize? He says he began getting the message right after buying a used ten-inch stone mill to grind some flour for his wife, Helen, to use in making bread.

Before long he was turning out small bags of flour under a "Natural Way Mill" label for health food stores in nearby Bemidji and Thief River Falls and for a church youth group selling it as a fund-raising project. The word spread almost overnight, he recalled, and he has been overwhelmed with business ever since.

"No one has ever shown this kind of interest in anything we've ever done before," he said. "All I know is that the visitors, the mail, and the phone calls would indicate nearly everybody wants some."

Although the Juhl farm is ten miles out in the country in the northwest corner of Minnesota, and only forty miles from the Canadian border, people from as far away as Minneapolis and St. Paul drive up on weekends to buy organic flour.

Juhl explained that these are individuals who stop by the farm to pick up twenty-five to fifty pounds of flour to take home. Others, he said, just send a letter with some money in it and say "send me some flour."

"There has been somebody stopping by almost every day, and every Sunday for sure, and we've been trying to get a sign up so people will know when there are tours," Mrs. Juhl added. "It's enjoyable but we never know when they are coming."

Although they have never advertised, the Juhls get dozens of phone and mail inquiries from bakeries, food co-ops, and other volume buyers looking for stone-ground flour. Most of them request samples and that involves putting a lot of flour in the mail.

Juhl is the kind of man who responds to a challenge. When he couldn't get rail cars three years ago to ship his grain to market, he bought a semi-trailer truck. His neighbors wanted some hauled, too, so he bought another. The result is a trucking operation, with three semis and several trailers dispatched regularly from the farm.

The trucks, with their regular runs to far-off places like Florida and California, fit right in with the expanding milling business. Juhl uses them both to deliver flour and other grain products to customers and to bring back organic corn and other grains for the mill.

Although he hasn't had much time to line up a steady supply of grain, he reported, there apparently is a substantial amount of organic wheat in neighboring North Dakota. This is hard red winter wheat, like he raises on his farm, and is the type used in making bread.

"We'll have to go to organic farmers in Iowa to get the corn we need and are prepared to reach out as far as we have to in getting the other grains," he said. "Of course, we raise quite a bit ourselves."

Juhl said he has started raising some hull-less barley, a high-protein cereal variety, for production of stone-milled barley flour. He said there appears to be quite a bit of organic rye and buckwheat in the region. He also has tried making flour from brown rice.

A rodent-proof building 40 feet wide and 104 feet long houses the new equipment, including two thirty-inch stone mills that turn out a ton of flour or more an hour. It also has storage space for at least three semiloads of finished product. Other equipment inside includes a high-capacity oats roller, grain cleaners, and hopper bins for cleaned grain.

Grain is brought through one wall into the cleaners from four 2,500-bushel hopper bins alongside the building. Juhl said the bins, plus other storage on the farm, provide enough capacity for between 60,000 and 70,000 bushels of grain of all kinds.

Automatic equipment fills bags to the desired weight and most will carry the "Natural Way Mill" label. It is likely, Juhl added, that printing on the bags soon will include some of his wife's good bread recipes.

Juhl has been farming without chemicals for three years and applies a humate-type organic fertilizer. The soil is high in organic matter, has not been cropped very long, and its main problem is that it often is too wet in the spring.

"Our yields have been comparable to those of others in the area," Juhl reported. "I think we're doing fine with organic methods."

For many years a large organic garden has supplied most of the food for the Juhl family. They no longer have livestock and prefer a diet of fruits, vegetables, grains, and nuts.

"We've been canning for five months, have three freezers full of food, and are ready for the winter," Mrs. Juhl said. "We practically live on what we raise here on the farm."

Nebraska's Organic Farmers

Many farmers have recently switched to "organic" or "natural" farming—that is, using no inorganic chemicals—and more farmers are becoming interested. Too little is known about why there is such a revival of interest in farming without chemicals. That is partly because of official skepticism and some downright scoffing. Yet natural farmers are a legitimate part of the farming community, and an increasingly interesting part. To find out why they are making the change, how they are doing, and how they view themselves, the Center for Rural Affairs informally surveyed 140 Nebraska commercial farmers reportedly using no inorganic chemicals on a substantial portion of their farm.

Below is a summary of the forty-two responses to the survey. In an attempt to express the beliefs and feelings of the farmers, their own words are used whenever possible.

Most of the farmers in the survey became interested in natural methods only recently. Twenty-nine of the forty-two stopped using chemicals within the last five years. Thirteen started farming naturally in the last year.

According to the farmers surveyed, one's decision to try natural methods comes after thoroughly examining all the aspects of raising crops and livestock. Farmers look at their chemical operations and find that many dismaying characteristics have developed over the years. Their soil has become continuously harder to work and less productive. Herbicides and pesticides, though generously applied, frequently are not controlling the weeds and pests. Fertilizers, increasingly expensive, are subject to shortages. The once-teeming wildlife has all but disappeared from their fields. The men find that farming, once the most peaceful of professions, has become deadly chemical warfare, and that they are on the hazardous front lines handling the highly toxic weapons of the war.

These farmers feel that something is drastically wrong with their farming operation; that somewhere along the line, they have gotten on the wrong track. Dennis Sanne of Clearwater echoes many of the farmer's sentiments: "I feel the time has come for. farmers to take a long look at the damage we are doing to our land and our lives with all the poisons we are using. The soil is a living thing we are very much dependent on, but we just treat it like dirt. We think only of how many dollars we can pull out of it without thinking of returning anything beneficial to it. Chemicals don't build, they only destroy."

The destruction of the soil is not the only concern. The farmers, aware of the toxicity of agricultural chemicals, worry about the consequences of poisonous residues in the grains they produce. Can those poisons, they wonder, be causing serious diseases in both livestock and humans? Some farmers feel there is a relationship between the increase in the use of agricultural chemicals and the increase of cancer in man. "I am not only concerned for my children's health," states Wayne Weiss of Boelus, "but for all the young people around the world. I feel that sickness among young people will be rampant not too many years hence. Our health is no better than the food we eat and the food is no better than the soil it is grown on, and the only good soil is that which man hasn't destroyed by chemical farming."

After making the decision to try natural methods, a farmer is confronted with the reality of raising crops without the chemical aids he has been so dependent on. Many of the farmers choose to start on only part of their fields, experimenting and learning from experience. However, some, convinced that chemical usage is wrong, one year stop using chemicals altogether, accepting the fact that their production could be down for a few years. In either case, making the adjustment to natural farming can be difficult. "I would say you might have to let natural farming grow on the man as well as the land for some of us who might not like to rotate as often as we should," states Jerome Noecker, a first-year natural farmer from Hartington. The more experienced natural farmers report that it takes a few seasons before a farmer really becomes proficient with the methods, and his soil begins its natural balance.

The typical natural operation surveyed consists of corn, oats, soybeans, and alfalfa grown in a strict rotation program. The purpose of the rotation is to keep the soil fertile and productive indefinitely without applying any oil-based synthetic fertilizers. There have been developments in the science of designing crop rotations and, as a result, farmers now have new information to help them plant crops in the proper sequence. To overcome weed problems the farmers plant later in the year. This has the twofold benefit of allowing them time to disc weeds under twice before sowing, and gives the crops a chance to shoot up rapidly in the warm soil. During the growing season, according to Del Akerlund of Valley, weeds are "taken care of through meticulous types of cultivation, rotation, and some hand care."

Not surprisingly, most of the farmers claim their input costs are lower with natural methods. What is surprising is that many farmers report that their yields without chemicals are comparable to their yields when the crop has the whole chemical treatment (the thirteen farmers who have used natural methods for only one year are unable to compare yields because of the unusual crop conditions in 1974 brought on by the drought). In fact, some farmers claim their yields with naturally grown crops are better than they have ever been when chemicals are used. This is especially true in the case of soybeans, oats, alfalfa, and wheat. With corn, most farmers acknowledge that the yields are usually less than chemically raised corn. However, they qualify that by saying the difference between the yields are minimal and that the natural corn has superior quality. They also report that in dry years, like 1974, corn grown naturally far outyields chemically raised corn.

With lower input costs and yields comparable to chemically raised crops, the farmers are able to realize a greater net profit. Contrary to the popular opinion that organic farmers can't succeed unless they can sell their products to "health food" stores for exorbitant prices, these organic farmers are succeeding, and even prospering, though they market their products through the same channels as other farmers.

Though many farmers report a greater net profit, this is not what seems to excite them most about natural farming. What is most commonly claimed as an advantage with natural methods is the improved quality of the soil. "The soil smells right again like real dirt used to," writes Richard Ruskamp of Dodge. "The ground is more mellow. Heavy machinery pulls easier. There is less erosion," writes Richard Heimer, a natural farmer for four years from Shelby. Dennis and Margaret DeBlauw of Fordyce, five-year natural farmers, claim there is "more humus in the soil. The soil is looser, easier to work, and it absorbs more moisture." John Rasmus of West Point who has farmed naturally for twelve years puts it simply, "it is easier to farm when your soil has life."

Amazingly, "as the soil becomes better balanced

through new and better soil testing, weeds aren't the problem," states Marvin Ruenhall of Syracuse. Indeed, many farmers assert that as their soils become more mellow, their weed problems diminish. Several farmers state that excessively weedy fields seem to be attributable to unhealthy soil. "Weed seeds will not grow as well in mellow soil," states Richard Heimer. Further, it is noted that it is easier to cultivate what weeds there are because the soil is looser.

When the farmers stop using chemicals, they further gain the advantage that natural enemies of the pests, such as birds and ladybugs, gradually move in to destroy the eggs of the insects, or to eat the pests themselves. In addition, rotation keeps most pests from proliferating. As the Bergman brothers of Elgin state, "When you rotate crops, you don't have to worry about rootworms, cornbores, or any other pests." Several farmers observe that healthy plants do not attract insects, that insects are merely nature's way of eliminating unhealthy plants. Bud Pritschau of Ravenna notes, "a good healthy plant will not have a pest problem. In a dry year you will have a problem with grasshoppers but in a dry year you won't have a healthy plant either. This is nature's balance at work." Indeed, when a farmer quits using chemicals, he finds that nature's intricate system of checks and balances resumes; he finds that nature is again able to maintain a proper balance of the living things on the farm. Natural farmers seem to accept the loss of a few plants to pests as part of nature's way.

Another frequently mentioned advantage of natural farming is that the feed grains produced without chemicals are of a better quality than chemically grown grains. It is reported that grains mature faster and need less drying when grown naturally. Farmers claim that livestock prefer to eat organically raised grains and that the livestock have less disease problems when they have a chemical-free diet. Some farmers mention that since switching to natural methods, they almost never need the veterinarian. In addition, the feed value of natural grains is deemed superior. Many farmers note that they need less feed supplements, and some also claim it takes less naturally grown grain to fatten the livestock. The Bergman brothers note that their grain is of a higher protein content, and is higher in total digestible nutrients.

Other advantages to natural farming given are: greater self-sufficiency, better family health, more earthworms, less plant disease, and, according to Ben Cuba of Silver Creek, "I can let my boys do any work there is to be done because there is nothing toxic."

In marked contrast to the many advantages that farmers list in their surveys, comments on disadvantages are few and far between. The most prevalent complaint is that there is more work involved, and that one has to be a better farm manager. One farmer notes that he needs more equipment for a rotation program than for monocropping. Another farmer says that natural methods can only be used on smaller operations, though the results of the survey do not bear this out. There are four families each farming 600 acres or more naturally, and the largest farm in the survey has over 1,000 acres of row crops, all raised without chemicals.

There are several other problems natural farmers have to contend with. Some farmers are disappointed that, because there is no special market, they have to take their grain to the local elevator and dump it in with all the chemically grown grain. As Ira Hunsberger of Ravenna states, "People around here couldn't care less how the food is grown." Another serious drawback, especially for the first-year natural farmers, is the lack of helpful, authoritative information on the unique problems that crop up when they switch from chemical farming. Steve Groetke of North Bend elaborates: "It is a different type of farming than we have been used to and it takes some adjusting and learning. There are very few places you can acquire information and help with natural farming procedures."

A further problem is the negative attitude of extension agents and neighbors, which makes it more difficult to get helpful information on the one hand, and on the other, causes them to be treated, as one farmer puts it, with a "certain amount of scorn." Farmers feel that natural farming has a long way to go toward winning solid support. A farmer from Howells states, "People get brainwashed by expensive chemical ads, TV and radio reports, university professors, and extension agents." Several farmers mention that their neighbors laugh at them, but one fellow remains philosophical about those who ridicule him in the face of evidence that he is successfully farming organically. Felix Cuba states, "the fertilizer dealers are not as friendly as they used to be—who cares."

Yet in spite of the negative reactions, many farmers report that their neighbors are becoming interested as they witness the method's success. Several Nebraska organic farmers have received nationwide exposure, and have been literally inundated with calls, letters, and requests for information.

Beyond the practical concerns, what truly sets natural farmers apart is their spiritual commitment to their way of farming. They feel that farming naturally means they are following God's plan: that they are no longer disrupting nature's way, but existing in harmony with it by nurturing their soil and animals rather than poisoning them. "God will bless our soil, water, crops, in fact our whole farming operation, provided that we are obedient to His laws," states Richard Miller of Elba who has been farming organically for four years. "If we fight nature," he continues, "which we are doing when we farm with chemicals, we are fighting a losing battle." Asserts George Myers, a natural orchard owner near Gibbon, "God desires that man live with His creation and not with contemptuous disregard for it."

The result of farming with "contemptuous disregard" for God's laws, many farmers feel, is that there will be a penalty to pay. "I feel the soil was given to man by God to maintain in its natural state. Everything that we eat comes from the soil in one way or another. If we upset the nature of the soil by using chemicals, we will hurt ourselves in the end," writes Gary Heine of Fordyce.

Because of the farmers' feelings that they are working in tune with nature, accepting the rhythms of nature, they appear to feel a contentedness, a "rightness" about their way of farming. They switch from chemical farming not for reasons of financial gain, but because they are disturbed by the lack of wildlife in their fields, or their ever-harder soil, or the residues of toxic chemicals they fear ingesting; they feel there is something "not right" with a way of farming that causes these problems. The decision to go with natural farming is for many a decision about the quality of life they wish to lead. As Del Ackerlund states, "it is a better way of life, especially being satisfied that we are not destroying or hurting any of the environment. I hope that some day soon the natural way of farming will take over completely."

Reprinted from the Winter, 1975 edition of the New Land Review, a publication of the Center For Rural Affairs; Walthill, Nebraska.

Organic Farming

One Grower's Move to a Saner Method of Farming

A young farmer like Steve Pavich is making waves in agribusiness. He's doing what most farm experts say can't be done: using more and more organic methods, applying less and less chemicals, improving crop production and quality.

On 600 acres of vineyards in the Delano, California, "grape belt," Pavich exemplifies the dramatic shifts that are taking place in American agriculture today. And interestingly, the message is being communicated through "official" channels. For example, the *Grape Grower*, a publication for large-scale commercial farmers in the West, featured Steve's methods in an October 1975 story headlined: "Better Land Management Key to Improved Grapes." The magazine explained that "Pavich now utilizes a composting manure program and has abandoned his insecticide spraying program entirely. He said he has not applied any insecticides on the Pavich vineyards for the past two years."

Five years ago, after graduating from California State University in Fresno with a degree in viticulture, Pavich started his farming career. By his standards, he was growing a poor-quality table grape.

He applied limestone as a way to improve water penetration, and it seemed to do the job. About two tons of limestone per acre are now spread and disced in twice a year. "We found that there was very little life in the soil. The earthworms, microorganisms, and predatory insects had either been killed by the chemicals or had left," he said. "We began experiments using different substances to control and promote changes in the soil balance. We started searching for alternative methods that might also increase production and quality if possible."

*For Steve Pavich, growing grapes organically has proven
to be a challenge. He has not applied any insecticides for
two years, and has not noted the insect problems everyone
said he would have.*

At first, some of the changes didn't work. "Soil
tests showed that the soil was low in potassium, so
we added potassium sulfate. It resulted in waterber-
ries. This indicated that in all probability we were
adding something that the soil was already producing to compensate for its needs. The potassium we
added pushed the soil out of balance in the opposite
direction. This was our last attempt to use commercial fertilizer."

Against advice from industry experts, he started
growing a cover crop in every other row. The winter
cover, planted in late fall, consists of 80 percent
vetch and 20 percent oats and is disced in as green
manure. Steve says he found the soil was much more
pliable in the rows planted to cover crops, so the
next season he began the practice in every row.

Franco Guerri, another grower, comments: "He has the best-looking plants and grapes anywhere. That soil is really soft—boy, it's just beautiful!"

Composted livestock manure is also put on the vineyard rows at a rate of approximately two tons per acre. In his first year of soil rejuvenation, Pavich applied five to seven tons to the acre, depending on soil tests. "If you can promote enough natural humus into the soil by growing it, adding it in compost, and making sure enough limestone is present, then the problems of low production and low quality seem to disappear," maintains Steve.

To lower the highly alkaline well water used for irrigation (pH 8.6), he also worked out a system for adding carbon dioxide, which drops the water pH to 8.6. Where severe alkali soil conditions and water pH are not limiting factors, it would be unnecessary.

As for the bugs, Pavich says the problem is now a minor one. "Ninety percent of the biological control I practice is by way of the soil," he insists. "That's the real answer—and it works. Chemicals treat symptoms, not causes," he adds. "When I had just finished college, I had never even heard biological control mentioned as an alternative. About the same time we were spraying for a severe case of leafhoppers and mites. On the third spraying of the season with our concentrate rig, our driver started complaining about being nauseated. So I decided to try myself since I couldn't believe any harm could be coming from the spray. I got the same result, and we started wondering if there wasn't a better way."

For the two worst insect pests of grapes—Pacific spider mites and grape leafhoppers—Pavich relies on a pair of natural controls along with the soil buildup. The mites get dusted with one-half ton of limestone per acre, which keeps the environment unfavorable for them, and their numbers at a minimum. Leafhopper infestations are curbed by introducing a helpful, tiny parasite, the *Anagerus* wasp, which feeds on the pests' eggs. He's also brought ladybugs, lacewings, and Chinese mantids into the action, and says they do stay around when no spraying disrupts them.

Sure, there are insects in his vineyards, Steve readily admits. "Look here, there are all kinds of insects. You go into most vineyards and there isn't any insect life on the surface or in the ground. He says if growers wouldn't kill all the insects they'll fight it out amongst themselves and the good bugs will take care of the bad bugs.

"It seems unhealthy vineyards attract mites and leafhoppers. Unhealthy plants may have an abundance of certain nitrates, which attract the insects. We also use ground kelp weed to help control insects and act as a fertilizer. Seaborn seems to be the most effective of the varieties we've tried."

As for weeds, let them coexist, says Pavich. "It's got to be destructive to call for eradication of plant life," he believes. "We still use cultivation to work up the soil, but we keep it to a minimum. The soil is left covered until early spring just before the bud-opening stage, then it is disced and the natural grasses and weeds are allowed to regrow. We also try to clip the grass and weeds every two weeks. When you kill the weeds and grasses, you kill the natural humus and aeration that promote water transfer. Certain of these weeds are also hosts to beneficial insects."

For effective pollination, Pavich has an area bee-keeper release bees every season. "He always brings more than contracted for." Steve smiles, "because my vineyards are unsprayed and the bees have a field day enjoying them." Windrowed wastes, mostly some 1,000 tons of dairy manure he composts, stand near the fields, ready for the one and one-half to two tons per acre he spreads each year.

Table grapes, Steve Pavich explains, are the hardest type to grow well; fruit for wine or raisins is easier. Yet 90 percent of his total production—600 to 700 boxes to the acre—make the quality grade. Gleaning black BIBIER varieties, red EMPERORS, amber ITALIA MUSCATS, golden THOMPSON SEEDLESS, and CALMERIAS are some of the best and tastiest grown in the West. The label on the lugs (and around the stems of grape bunches) says it all: "These Sunshine California Grapes have been grown in limestone soil without the usage of toxic insecticides, or chemical fertilizers in hope that people can appreciate a natural good taste."

Pavich employs 25 to 50 steady workers, up to 150 during harvest. Most of his crop goes to wholesale brokers at terminal markets and to a few national chain markets. His investment represents about $1,500 an acre, he told me. One idea Steve has is to start stores in the Los Angeles area, especially near UCLA, run by farmers. He's also active in, and enthusiastic about, the California Certified Organic Farmers group. A goal he has in mind is to form a co-op of concerned farmers and provide a central source of supplies and reliable information for biological growing. Growers representing some 10,000 acres are now working on its organization to reduce costs and increase quality production.

Chapter 2

The Economics of Organic Farming

This chapter is adapted from a report prepared by the Center for the Biology of Natural Systems, of Washington University. This two-year study is the most thorough research done to date on the economic feasibility and energy consumption of organic farming. Since farming is a highly individual pursuit with uncountable variables, many of the scientific methods used in the report leave room for argument. The authors of the report are aware of this, and say only that their work is the most accurate done in this field at this time. The project has been well received by the traditional agricultural press. The report is a product of the efforts of William Lockeretz, Robert Klepper, Barry Commoner, Michael Gertler, Sarah Fast, Daniel O'Leary, and Roger Blobaum. This edited version is solely the responsibility of the editors of *Organic Farming: Yesterday's and Tomorrow's Agriculture*. Copies of the original report (Organic and Conventional Crop Production in the Corn Belt: A Comparison of Economic Performance and Energy Use for Selected Farms, CBNS-AE-7) is available from the Center for the Biology of Natural Systems, Washington University, Box 1126, St. Louis, Missouri 63130.

In the last twenty-five years there have been major changes in United States agricultural methods. Perhaps the most striking change has been the rapid growth—almost a five-fold increase since 1950—in the use of agricultural chemicals, especially inorganic fertilizers and synthetic pesticides. In just five years (1966–1971) the quantity of active pesticide ingredients used annually on crops rose by 40 percent to 466 million pounds.

These changes have been particularly striking in the Corn Belt. Before World War II, almost no inorganic nitrogen fertilizer was used and modern herbicides and insecticides (especially the chlorinated hydrocarbons) were not yet on the market. Now inorganic nitrogen fertilizer is commonly applied on corn land at rates well in excess of 100 pounds of nitrogen (N) per acre and synthetic pesticides are widely and regularly used.

These new features of agricultural technology are credited with most of the post-war rise in the productivity of Corn Belt agriculture, such as the doubling of the yield per acre of corn since 1950. At the same time, the intensive use of such agricultural chemicals has played an increasingly important role in the economics of Corn Belt agriculture. The cost of fertilizers and pesticides now constitutes approximately one-half of the variable costs of producing corn. These inputs have yielded particularly high economic returns. For example, the increased yield of corn in Illinois, at a typical level of nitrogen fertilization, compared to the yield when no fertilizer or manure is applied, is worth five to ten times as much as the cost of the fertilizer. Such financial returns are the basis of the widespread conviction that the high productivity of U.S. agriculture and the economic well-being of the U.S. farmer depend crucially on the present level of use of fertilizers and pesticides.

Despite their acknowledged advantages, some questions have arisen regarding the wisdom of continuing these practices, in particular the heavy application of nitrogen fertilizer and the regular use of synthetic herbicides and pesticides. These questions arise from two quite diverse origins: the energy crisis and ecological concerns.

With the advent of the energy crisis, farmers have been faced with threatened shortages and sharply increased prices of fertilizers and pesticides, since their manufacture requires petroleum and natural gas as a source of both the raw materials and the heat used in the manufacturing processes. For example, in Illinois, the price of 100 pounds of anhydrous ammonia fertilizer (the most common nitrogen fertilizer and a major input in corn production,) increased from an average price of $4.30 in April 1973 to $14.25 in April 1975. Further increases in the price of nitrogen fertilizer and other energy-intensive agricultural chemicals are a likely consequence of future energy price increases. Since these inputs represent an appreciable part of the Corn Belt farmers' production costs, continued price increases threaten to seriously reduce farm income, especially if crop prices decline.

A second source of questions regarding the advisability of continuing the present heavy reliance on agricultural chemicals is the possible adverse environmental effects that may result from the use of these chemicals. Examples of such effects include inadvertent harm to humans or wildlife from toxic pesticides, or excessive levels of nitrate in ground water and streams from fertilizer nitrogen draining from agricultural lands.

Against this background, one can see why a farmer might be interested in the possibility of avoiding the use of fertilizers and synthetic pesticides. Most commonly referred to as organic farming, but also known by other names such as "natural farming" and "biological agriculture," this method requires that plant nutrients be supplied in one or more of the following forms: organic wastes, most commonly livestock manures; leguminous green manures; purchased organic fertilizer materials; and inorganic minerals, such as rock phosphate, that have not been chemically treated. On Corn Belt organic farms that raise field crops, weeds are controlled mechanically and by selecting crop sequences intended to minimize weed problems, such as alternating meadow with row crops every few years. The main means of insect control is to use crop rotations chosen in part to reduce pest problems, such as avoiding corn after corn. (Insect problems in other agricultural systems, such as fruit production in California, are often more severe than for field crops in the Corn Belt. In such cases, organic farmers may use additional biological control techniques).

There is, at this time, no general agreement concerning what practices should be included under the term "organic farming." In this study, we use it to denote farming in which nutrients are supplied to the soil in an organic form such as livestock manure or green manure crops, but with no use of inorganic nitrogen, phosphorus, or potassium fertilizers. Also, no pesticides (herbicides or insecticides) are used. In contrast, we use the term "conventional" to refer to farms on which inorganic fertilizers and pesticides are used. These farms may also use livestock manures. (In fact, since all the conventional farms described in this report raise livestock as well as crops, they all have manure available, although, as will be seen later, it is used to a somewhat lesser degree than on the organic farms we have studied).

In discussions concerning organic farming, a frequent source of confusion is the fact that the word "organic" is also used to classify substances chemically. An organic chemical is one in which the element carbon, linked together in straight or branched chains, or rings of various sizes, forms the central core of the substance's molecular structure. "Organic farming" is not at all identical to "farming with organic chemicals": for example, chlorinated hydrocarbon insecticides such as aldrin and dieldrin are organic chemicals, but are never used by organic farmers. On the other hand, organic farming principles do require that plant nutrients be provided in organic form, even though plants always take up nutrients in inorganic form, such as nitrate, phosphate, etc. The reason for this practice is that when organic materials such as manure are broken down in the soil to yield inorganic plant nutrients, they also provide organic carbon compounds that can be incorporated into the soil's reserve of organic matter (humus). Because of the generally favorable effects of soil organic matter on soil texture, water retention, and "tilth," an increase in soil organic matter levels is an important goal of organic farming. (Whether this goal is actually achieved depends on how much manure or other organic matter is applied, what crops are raised, their yields, etc., so that no general statement can be made concerning how organic and conventional practices compare in this regard.)

A small minority of farmers in the Corn Belt have made this choice. They have completely eliminated their dependence on inorganic fertilizers, herbicides, and insecticides. To provide plant nutrients, they use livestock manures and crop rotations involving nitrogen-fixing legumes, such as clover. In some cases, commercial organic fertilizer materials are applied as a supplementary nutrient source. To control weeds they use more frequent cultivation and rotate row crops with hay, which is mowed frequently to clip weeds. They generally make no special effort to control insect pests, except by avoiding certain sequences of crops, particularly corn after corn. In most other respects, however, these organic farms are similar to conventional Corn Belt farms. Mechanized methods are used almost exclusively, with a small amount of hand-weeding of soybeans. Their usual size is a few hundred acres, which is typical of commercial farms in the Corn Belt, and they raise the main crops of the Corn Belt: corn, soybeans, small grains, and hay. In most cases, their products are sold in the regular agricultural market at prevailing prices, although some are sold specifically to the "organic foods" market.

Obviously, in choosing this course, the farmer is

willing to give up the considerable economic advantages which, it is believed, may be gained by using these inputs. He must therefore believe, on the contrary, that an economically successful operation is possible without them.

To determine whether or not the intensive use of pesticides and inorganic fertilizers is essential to the success of U.S. agriculture, at least as represented by certain kinds of Corn Belt farms, we have undertaken to compare two samples of Corn Belt farms: the first is a group of organic farms on which no inorganic fertilizers or insecticides, and almost no herbicides, are used; as a control sample, we have studied a group that is otherwise quite similar, but which follows conventional fertilization and pest control practices. The goal has been to compare these groups with respect to the following measures of performance:

1) Total market value of crops produced per acre;
2) Economic returns from crop production (value of production less costs);
3) Energy-intensiveness, or quantity of energy consumed in crop production divided by the value of production.

The second of these measures is important from the viewpoint of the farmer himself since it determines whether a farm is economically viable. In contrast, the first and third items reflect certain social consequences of the two systems. Total value of production gives an indication of the overall level of output of each farm.

For the past two years, we have been studying the economic performance and energy consumption of a group of Corn Belt farms that use organic methods, as defined above. A previous report described the results of the first year of this study, which concerned the 1974 crop year. The basic findings of that report were: 1) the organic farmers' gross revenue from crop production per acre was 8 percent below that of a matched group of farmers who used chemicals; 2) the returns to crop production, i.e., gross income less operating expenses, were virtually identical in the two groups; 3) the energy-intensiveness of the organic group was slightly more than one-third that of the comparison group.

The organic method has long been dismissed as not deserving serious consideration as an alternative to conventional commercial-scale agricultural practice. Many people believe that it is hopelessly non-competitive, and that the success of Corn Belt and other commercial-scale agriculture is virtually impossible without the use of inorganic fertilizers and synthetic pesticides. If, on the other hand, organic farms appear to be close to conventional farms in economic returns, as indicated in our first report, then the system suggests a direction towards which farmers might wish to move, especially if fertilizer prices continue to rise and the supply situation continues to be tight, or if crop prices fall relative to the prices of fertilizers and pesticides.

Because the application of organic matter to the soil is an intrinsic part of organic farming, all the farms in the organic sample raise livestock, which serves as a source of manure. Therefore, to provide a valid comparison, we chose matching conventional farms that also produce livestock. In both the organic and conventional groups, crop production is dominated by livestock feeds—corn, soybeans, and hay—along with some production of small grains (especially wheat) for human consumption. In this respect both groups are representative of Corn Belt agriculture, which is a major producer of hogs and fat cattle, as well as of the crops that are fed to these animals.

We studied fourteen organic farms operated as family enterprises by full-time farmers: four each in Illinois and Iowa and two each in southern Minnesota, northern Missouri, and eastern Nebraska. All of these farms were studied for 1974 and 1975. Two additional organic farms and their matched conventional farms that had been studied in 1974 were not included in the 1975 study.

All of these farms have one or more forms of livestock, either a beef-cow herd, cattle on feed, hogs, or a dairy herd. At the start of the study their average total size was 429 acres. Their average amount of cropland (as defined in more detail later) was 250 acres, with a range from 102 to 420 acres. Four of the farmers rented additional cropland on which chemicals were used, but such land is not included in the study.

No inorganic nitrogen fertilizer or urea, acidulated phosphates (e.g., triple superphosphate), or—with one exception—conventional potassium fertilizer (muriate of potash) were used on the land included in this analysis. Some of the farmers use commercial organic fertilizers, phosphate rock, or powdered rock containing trace minerals. No insecticides are used on any of these farms, and in all but one case no herbicides are used. In all cases but one, these farms once had been operated using inorganic fertilizers and/or pesticides. They have all been operated organically at least since 1970; the median date of conversion to organic management was 1968.

These farms were located for this study primarily by word-of-mouth; to be included we required each one to be within the range of 175 to 800 acres total size, to have been operated organically at least since 1970, and to produce both field crops and livestock.

In addition, they were chosen to be widely distributed over the central and western portion of the Corn Belt.

TABLE 2-1

COMPARISON OF THE VALUE, SIZE, LIVE-STOCK INVENTORIES, AND EQUIPMENT OWNED BY THE TWO SAMPLES OF FARMS IN 1974.

Characteristic	Organic	Conventional
Average of farmer's estimates of land value (dollars per acre)	836	841
Average total size in acres	476	462
Average cropland in acres	266	358
Average livestock inventory (animal units)*	137	128
Percent owning tractor	100%	100%
average size of largest	95 hp	96 hp
Percent owning row crop planter		
at least 4 rows	100%	100%
larger than 4 rows	25%	12%
Percent owning self-propelled	68%	68%
combine with cornhead	63%	53%
with cornhead larger than 2 row	32%	26%
Percent owning moldboard plow	25%	94%
average width of largest (bottom width times no. of shares)	72″	72″
Percent owning chisel plow	94%	35%
average width of largest	11.3″	10.8″
Percent owning at least two grain		
trucks or wagons	94%	100%
average capacity of largest	245 bu.	230 bu.

* Excluding one pair in which the organic farmer had sold off his cattle the previous year.

This group of farms is best regarded as constituting a set of case studies, rather than a randomly selected statistical sample. Because of the sampling procedure we used, our results should not be extrapolated to the population of all organic farms in the Corn Belt. Table 2-1 profiles the farms studied.

Each organic farm was paired with a nearby conventional farm that was roughly similar in soil types, size, and livestock inventories. Their average total size at the start of the study was 479 acres, with an average of 348 acres of cropland (range: 124 to 769 acres). Each of the conventional farmers was individually recommended by staff members of his county office of the Agricultural Stabilization and Conservation Service as being a "top management" operator who used modern production methods.

Fertilizer use on these farms is shown in Table 2-2; this shows that this group of farmers used fertilizer application rates that on average are very close to those used by all farmers in their areas. All of these farmers used some herbicides or insecticides. On corn, thirteen of the fourteen farmers used herbicides and ten used insecticides; on soybeans, eleven used herbicides. Other use of pesticides was minor.

TABLE 2-2
Average fertilization rates, conventional sample and all farms in some states, 1974. (All rates are pounds N, P_2O_5, and K_2O per acre receiving fertilizer.)

Crop	Conventional Sample	All Farms
Corn	101-53-71	108-61-63
Soybeans	8-38-40	13-44-66
Wheat	53-67-57	55-61-53

To get some measure of the extent to which the organic farms and their conventional matches had cropland of comparable productivity, we obtained at least two estimates of the rental value of each farm's cropland from county ASCS officials and other experts on local agricultural conditions, such as extension agents and Federal Land Bank officials. These estimates are summarized in Table 2-3. The table shows that on the average, the two groups had cropland of roughly comparable value, although within a specific pair there were considerable discrepancies.

TABLE 2-3
Estimated cash rental value of cropland per acre, 1975.

	Organic	Conventional
Mean	$68	$66
Range	$34 to $99	$39 to $104

The matching procedures we have used are at best only approximate, and still leave room for considerable within-pair differences, particularly in matching soil types. The farms in a pair are always at least in the same soil association, but do not necessarily have the same types within the association.

For each farm, the following data were collected for each field:

1) Crop(s) and yield(s);
2) Application rates and costs of all fertilizers, pesticides, lime, and other materials;
3) Types of manure and application rates;
4) Tillage, cultivation, planting, and harvesting operation;
5) Seed variety, price, and seeding rate.

An important difference in the data collection between the two years was that whereas for our study of the 1974 crop year we interviewed each farmer during the winter of 1974-75, for 1975 we provided each farmer with data forms before planting began in the spring. These forms, which covered all preharvest operations, were returned to us during the summer, after which we followed-up by telephone or letters if necessary to correct omissions, misunderstandings, etc. A second set of data forms covering the harvest period was sent out in the summer, and returned to us in late fall or early winter. This method overcame the problem we had in the 1974 data collection, which required that the farmers provide us with information on what they had done the previous season, without our having informed them in advance as to the kinds of records they should keep.

All analyses presented in this report concern only the crop production on each farm, although all the farms in this study also raise livestock. We cover all operations from primary tillage to the hauling of the crop to an on-farm storage site, including whatever operations are needed to make it ready for storage (e.g., corn drying). Manure hauling and spreading are considered part of crop production.

Because of the differing rotations used in the two systems, the two groups of farms differ in the proportion of cropland devoted to various crops, even though the main crops raised on each are the same. Consequently, we present all crop production data in the form of averages over all cropland on each farm. *Cropland,* as we use the term, means all land that was ever in crops during the preceding five years, regardless of how it was actually used in the 1974 and 1975 crop years. That is, it includes: land in row crops or small grains; land in cultivated pasture and/or hay, if it is in rotation with other crops;

land temporarily in soil improvement crops; and land unharvested in either 1974 or 1975 for any reason, but harvested in any of the three previous years. The only land not included is permanent pasture or permanent hay land (defined as land that has been in pasture or hay for the last five consecutive years), as well as feedlots, woodlots, land associated with buildings and facilities, etc.

The organic group had ninety-eight acres less cropland than the conventional farms on the average, although their average total size was only fifty acres less. One reason for this is that some of the organic farmers also farm additional cropland with chemicals, this land is not included in the study.

A second factor is that the organic group has more permanent pasture, and also has more animals fed primarily on pasture (cows, calves, and bulls): an average of 101 animal units, compared to 63 for the conventional group. Animal units were calculated assuming cow = 1.0, calf = .3, bull = 1.4, feeder cattle = .7, and hog = .2.

Unfortunately, many of the farms in this study have not been mapped for soil type. Consequently we cannot say whether the difference in the pasture/cropland ratio on the two groups of farms represents a difference in the way the farmers choose to manage land with a given set of characteristics, or whether it reflects an actual difference in the proportions of various types of land on the two groups of farms.

Although the organic farmers make use of leguminous green manures, those in our study sample all do so in a way that permits them to obtain at least some harvested crop during a one-year cycle. There are three ways of doing this: 1) a winter-hardy legume is seeded into corn in the fall, and plowed under the following spring; 2) a stand of leguminous hay is cut only once (or twice), with subsequent growth plowed under; 3) a legume is seeded with a small grain, which is combined in the early summer, with the legume being plowed under the following spring. In all such cases, we included all costs associated with the green manure, without crediting it with any value.

Table 2-4 shows the machinery and equipment complements of both groups. Since the two groups are similar in equipment size, in computing labor requirements and operating expenses for field equipment, we assumed a uniform size for tractors, combines, implements, etc.

TABLE 2-4
SELECTED CHARACTERISTICS OF MACHINERY AND EQUIPMENT, 1975.

Item	Organic	Conventional
Tractor	89 hp 65 to 125 hp largest range	94 hp 55 to 125 hp largest range
Moldboard plow	29% owning 64" median width	86% owning 72" median width
Chisel plow	100% owning 11 feet width	50% owning 10 feet width
Cornhead for combine	57% owning	57% owning

In calculating energy consumption, we included the fuel used in all field operations and hauling to on-farm storage, in drying corn, and in manufacture of fertilizers, pesticides, and soil amendments. Field operations were assumed to be done by diesel;

manure spreading, combining, and hauling by gasoline.

CROPPING PATTERNS

Although the two groups of farmers raise the same major crops, they do so in different rotations, so that they differ in the proportion of their cropland in each crop in a given year. As seen from Table 2-5, corn (for grain or silage) is the leading crop on both groups of farms; however, because of the importance of hay and pasture in organic rotations, row crops (corn and soybeans) average only 52 percent of cropland in the organic farms compared to an average of 73 percent on the conventional farms. Small grains are more prevalent on the organic farms because these crops are generally raised as nurse crops for new stands of meadow.

On the organic farms, common rotations are: corn/soybeans/small grain/hay; corn/soybeans/ corn/small grain/hay; corn/small grain/hay; and corn/soybeans/small grain/hay/hay. Depending on which particular rotation is followed, the proportion of land in each crop will be within the following limits: corn: 20 percent to 40 percent; soybeans: 0 to 25 percent; small grain: 20 percent to 33 percent; and hay: 20 percent to 40 percent. The figures in Table 2-5 reflect a mix of these rotations, plus some less common variations. On the conventional farms, common patterns are: continuous corn; corn/soybeans; and corn/soybeans/small grain/hay. This leads to the predominance of row crops shown in the table. A slight amount of double cropping is reflected in those totals which are greater than 100 percent. The double cropping is almost always winter wheat followed by soybeans. (In the table, a new stand of hay established with a small-grain nurse crop is not

The Economics of Organic Farming

considered as double cropping, and is included only under the nurse crop.)

TABLE 2-5
AVERAGE FRACTION OF CROPLAND IN VARIOUS CROPS FOR 1975 AND 1974.

Crop	Organic		Conventional	
	1975	1974	1975	1974
Corn Grain	25%	24%	39%	35%
Corn Silage	6	8	4	6
Soybeans	21	21	31	31
Wheat	9	6	7	7
Oats	11	12	6	5
Hay or Pasture*	25	25	12	15
Other**	3	5	3	4
Total	100%	101%	102%	103%

*Established stands only.

**Includes milo, rye barley, buckwheat, and new meadow seedings other than with nurse crop.

YIELDS

Table 2-6 shows the yields obtained by the two groups of farms in 1974 and 1975. Because this table is intended to indicate how the two groups compared in yields, it is restricted to only those farms in which both members of a particular pair raised a particular crop. The clearest difference between the two groups is in corn for grain, for which the conventional group obtained higher yields. Differences in wheat (1975) and hay (1974 and 1975) are not particularly meaningful because there were so few farms included in the data.

Table 2-6 shows that, in general, yields improved in 1975, and that the conventional group's advantage over the organic group was stronger in 1975 than 1974. These are understandable consequences of the different weather patterns. In 1974, virtually every farm in our study was adversely affected by either an extremely wet spring, a severe drought during the summer, an unusually early frost, or a combination of these conditions. In 1975, some of the farms were also affected by drought, but weather conditions were good for a large portion of the samples. Consequently, yields were appreciably better. The better relative performance of conventional farms under these conditions may reflect the fact that when yields are limited by poor weather less is gained by increasing the nutrient supply with inorganic fertilizers.

TABLE 2-6
AVERAGE YIELDS FOR 1975 AND 1974

	Organic		Conventional		Number of Pairs Raising Crop	
	1975	1974	1975	1974	1975	1974
Corn Grain**	74	74	94	76	14	11
Soy-beans**	35	32	38	29	12	11
Wheat**	28	28	38	29	4	4
Oats	58	56	60	60	5	5
Hay***	4.5	5.0	3.9	3.4	4	6

The yields for four major crops obtained on all farms in our study, regardless of whether the matched farm in a pair raised the same crop, are shown in Table 2-7. In this case, the average yields of the two samples should not be compared directly to each other, since they reflect different portions of our study area. Rather, for each sample and each crop, we have compared the sample's average yield to the county-wide average yields in the counties in which there was a farmer in the sample who raised that crop. The conventional group did about as well as the general population in 1974, and considerably better in 1975. The organic group did noticeably better in soybeans in both years, but got much lower yields in corn and wheat than farmers in general in 1975. In all other cases, their yields were about the same.

TABLE 2-7
AVERAGE YIELDS FOR BOTH SAMPLES, AND ALL FARMS IN SAME COUNTIES.

Crops	Year	Organic Sample	County-wide Acreage	Conventional Sample	County-wide Average
Corn	1975	74	90	94	88
	1974	74	75	71	73
Soy-beans	1975	34	30	38	31
	1974	32	25	28	25
Wheat	1975	26	38	41	36
	1974	28	31	29	29
Oats	1975	56	NA	57	NA
	1974	59	55	59	59

Table 2-8 shows the average value of all crops produced on each type of farm, expressed as market value per acre of cropland. By weighing the value of each crop on a particular farm in proportion to the amount of cropland in that crop, we are able to compare gross income from crop production on farms with varying proportions of land in each crop. We use market value for all crops regardless of whether or not the crop was actually sold or was consumed by livestock on the same farm. This device was used as a way of separating crop production from other enterprises on the farm, since all of the farms in this study, conventional as well as organic, have livestock.

The higher value of production on the conventional farms reflects two factors: the higher yields, especially corn, and the higher proportion of cropland in high-value crops, especially corn and soybeans. The average value on the conventional farms was higher than on the organic farms in both years. The differential was more pronounced in 1975, which could be expected from the fact that the difference in yields was much greater that year. However, the conventional group's advantage in comparative market value in 1975 did not increase by as much as its comparative yields. This is because corn and soybean prices declined in 1975, whereas roughages became more valuable.

Data on the operating, or out-of-pocket, costs for both groups, averaged over all cropland, are shown in Table 2-8. Apart from a general inflation in input prices, there were no significant changes from 1974 to 1975 in the overall average level or range of operating costs in our two samples. The difference in production costs (nineteen dollars per acre averaged over the two years) results primarily from the fertilizers and pesticides purchased by the conventional farmers. Although the organic group bought some fertilizer materials, such as rock phosphate, these expenditures were quite small compared to fertilizer purchases by the conventional group. Other costs for a given crop were about the same. A small additional contribution to lower costs on the organic group comes from their having a greater proportion of their land in crops requiring fewer inputs, e.g., established hay relative to corn grain. As a fraction of total value of production, costs on the organic farms averaged 19 percent, compared to 27 percent on the conventional farms. If production costs rise more rapidly than crop prices, or if crop prices decline, as occurred between 1974 and 1975, the conventional group will be more seriously affected; in contrast, they will benefit more if crop prices rise more rapidly than production costs, as occurred between 1972 and 1973.

TABLE 2-8
AVERAGE MARKET VALUE OF CROPS PRO-
DUCED, OPERATING COSTS, AND CROP PRO-
DUCTION RETURNS, PER ACRE OF CROPLAND.*

Year	Market Value of Crops ($/A)		Operating Costs ($/A)		Crop Production Returns ($/A)	
	Organic	Conv.	Organic	Conv.	Organic	Conv.
1975	169	193	34	54	135	140
1974**	159	172	28	46	131	127
Two-year Average	164	183	31	50	133	133

The difference of value of crops produced, less operating costs, which we refer to as crop production returns, is also shown in Table 2-8. This quantity was almost equal for the two groups in both years. As with market value of crops produced, however, there is a considerable variation within each sample, particularly the organic group.

ENERGY INTENSIVENESS

To take into account the somewhat different production levels and crop mixes on the two kinds of farms, we compute energy intensiveness as energy consumed in crop production per unit market value of crop produced, expressed as BTU/$. As with costs, production, and returns, this quantity is averaged over all cropland in each farm.

The energy intensiveness of the two groups of farms in our study is shown in Table 2-9. In both years, the conventional group was an average of 2.3 to 2.4 times more energy-intensive, with the disparity somewhat more pronounced in 1974. For both groups, the energy intensiveness declined slightly in 1975. These slight changes reflect the higher levels of production in 1975, rather than a change towards less energy-intensive practices. That is, a major fraction of the total energy requirement is independent of yields, so that energy intensiveness declines when yields are higher. Likewise, the improvement in yields was greater for the conventional group, as we have already seen. Consequently, that group became slightly less energy-intensive relative to the organic group (but still was higher by a factor of 2.3), as well as in an absolute sense.

The main cause of the difference between the two groups comes from the use of inorganic fertilizers, especially nitrogen, by the conventional group. In terms of energy per unit of crop, the two-year average energy intensiveness for corn was 24.7 thou. BTU/bu. for the organic sample VS 68.4 for the conventional sample. For soybeans the difference was much less pronounced: 33.7 thou. BTU/bu. (organic) VS 39.4 (conventional). The difference between the two groups is smaller than for corn primarily because very little nitrogen fertilizer is used on soybeans.

We have compared the amount of labor used directly in crop production in 1975 on both groups of farms. As mentioned earlier, the two groups are similar in size of their equipment, so that this assumption simplifies the calculations but does not differentially affect the two groups. Likewise, since the two groups were roughly matched by type of soil, we have assumed that a given operation, when performed by our hypothesized standard equipment, takes the same length of time per acre on both groups of farms.

TABLE 2-9

ENERGY INTENSIVENESS OF CROP PRODUC-
TION (THOU. BTU PER $ OF MARKET VALUE.)

Year	Organic	Conventional
1975	6.6	15.0
1974	7.2	17.3
Two-year Average	6.9	16.2

The average labor requirement for the two groups is given in Table 2-10. For all cropland taken together, the organic group is 3 percent more labor-intensive in terms of labor per acre. However, because their level of production per acre is lower, the difference in labor intensiveness is more pronounced (12 percent) when expressed as labor input per unit value of crop produced.

TABLE 2-10

LABOR REQUIREMENTS FOR CROP
PRODUCTION 1975.

Crop	Organic	Conventional
All per acre per $1000 of production	3.3 19.8	3.2 17.8
Corn grain	3.9	3.8
Soybeans	3.1	2.6
Small grains	1.9	1.9

(Hours per acre, except where indicated.)

Table 2-10 also shows the labor requirements of major crops. For corn grain, soybeans, and small grains, the labor requirement shown in Table 2-10 for both groups is somewhat lower than the national averages for 1969–1973. Apart from the general reduction in labor requirement in the past few years, this difference also reflects the highly mechanized character of the farms in both samples, which is reflected in our calculations based on combines with four-row corn heads, a tractor of 95 hp, etc.

The breakdown of labor requirements by task, for corn grain, is given in Table 2-11. The biggest differences are in cultivation, which requires more labor on the organic farms because no herbicides are used, and in chemical application (dry fertilizer, anhydrous ammonia, herbicides, insecticides, and

TABLE 2-11
AVERAGE LABOR REQUIREMENTS IN HOURS PER ACRE 1975

Task	Organic	Conventional
Seedbed Preparation	.77	.85
Chemical Application	.08	.30
Manure Application	.52	.46
Planting	.30	.27
Cultivation	.53	.28
Harvesting and On-Farm Hauling	1.33	1.38
Shelling and Drying	.38	.28
TOTAL	3.91	3.92

soil amendments), which is more important on the conventional farms. These differences essentially offset each other; the total labor requirements differ by .1 hour per acre. However, this calculation, like the others reported above, includes all field operations, regardless of who performs them. Fertilizer and pesticides application is often done on a custom basis, which means that farmers in the conventional group can reduce their own expenditures of labor. However, to the extent that chemical application is done by custom operations the production costs given earlier should be increased. Prevailing custom rates are generally higher than the figures we have used, since we include only direct costs.

MANURE APPLICATIONS

Spreading of livestock manure on cropland is a common practice on both samples of farms. Table 2-12 shows the distribution of manure applications over various crops, as well as average application rates. For example, manure was applied to at least some soybean land on five of the thirteen organic farms on which soybeans were raised in 1975. On these five farms, an average of 56 percent of the soybean land received manure; on the soybean land receiving manure, the average application rate was 8.8 tons per acre.

For all cropland taken together, the two groups were fairly similar in the prevalence of manure spreading, the application rates, and the fraction of land receiving manure. However, there was a somewhat greater tendency on the part of the organic farmers to spread the manure over more crops, whereas the conventional farmers applied it mostly to corn.

These figures only refer to manure spread by the

TABLE 2-12

LIVESTOCK MANURE APPLICATIONS, 1975

Crop	Organic			Conventional		
	% of Farms	% of Land Receiving Manure	App. Rate (t/A)	% of Farms	% of Land Receiving Manure	App. Rate (t/A)
Corn for grain	50% (7 of 14)	64%	8.2	50% (7 of 14)	54%	9.1
Corn for silage	67% (4 of 6)	89	6.8	57% (4 of 7)	79	10.3
Soybeans	38% (5 of 13)	56	8.8	17% (2 of 12)	61	8.0
Hay	29% (4 of 14)	51	6.0	10% (1 of 10)	100	7.3
All Crops	71% (10 of 14)	28	8.4	64% (9 of 14)	28	9.3

farmer. Another source of manure for cropland is livestock grazing on temporary pasture, as well as animals that are being fed on cropland after harvesting.

NUTRIENT BALANCE

We have attempted to estimate the net balance of all inputs and removal of P and K from the cropland on both samples of farms. Nitrogen is not included in this analysis because of the tremendous uncertainties in knowing how much is fixed by legumes, the loss to the environment through leaching and denitrification, and the loss from manure during storage and application. In contrast, we have assumed no environmental losses for P and K, the major cause of which is erosion.

This computation takes into account inputs in the form of commercial fertilizer and livestock manures, and removal in the harvested portion of the crop. Lacking measurements of actual nutrient concentrations in crops harvested on our samples, we have assumed textbook average values. This simplification is more likely to introduce an error in the case of forage crops, which are highly variable according to growing conditions, time of harvesting, etc. For grains, there is also some variability, but it is less significant. We also used standard values of nutrient content of each type of livestock manure (dairy cows, beef cattle, hogs). We take into account the way in which the crop was harvested, e.g., corn grain (combined), ear corn, or corn silage. The only exception is for small-grain straw that is baled as animal bedding after combining. This straw was not considered part of crop production in computing energy, income, and costs, and likewise is not included as a nutrient drain.

As with income and costs, we express all results as averages over all cropland, rather than on a field-by-field basis. The results of these calculations are summarized in Table 2-13. Table 2-13 shows that the conventional farms are just about in equilibrium with respect to P and K. In contrast, the organic farms appear to be drawing on the reservoirs of P and K. The 41 pound per acre per year depletion of K_2O in large part results from the importance of hay in the organic rotations.

The calculations presented here have many simplifications. The results are averaged over all cropland, which means that there could be appreciable deficits on some fields and surpluses on others: over the long run, these average values would be realized on individual fields only if the fields were managed according to a strict rotation. Also, we make no distinction between the various degrees of availability that P and K can have. Likewise, we do not consider the depths from which P and K can be drawn (and

Organic Farming

TABLE 2-13
AVERAGE BALANCE OF P AND K ON OR-
GANIC AND CONVENTIONAL CROPLAND, 1975.

| Type of Farm | Nutrient | Inputs | | Removed | Net |
		Fertilizers*	Manure	In Crop	
Organic	P_2O_5	+ 6.9	+ 9.7	- 28.9	- 12.3
	K_2O	+ 4.6	+ 16.8	- 62.1	- 40.7
Conventional	P_2O_5	+ 23.4	+ 7.4	- 29.9	+ .9
	K_2O	+ 32.4	+ 12.5	- 43.8	- 1.1

*Lbs. P_2O_5 and K_2O per acre per year

the varying depths to which they can be returned in crop residues) when there is a sequence of crops of differing root depths. Thus our results are therefore a gross measure of the total input/output balance, intended to provide an approximate measure of the extent to which the two groups of farmers may be drawing on their soils' preexisting nutrient levels.

SUMMARY OF FINDINGS

The second year's data reported here are qualitatively compatible with the preliminary conclusions that we reached earlier on the basis of 1974 data only: 1) the organic group produces crops at a slightly lower overall level, as measured by market value per acre; 2) profitability of crop production per acre of cropland is comparable on the two groups of farms; 3) the organic group consumes appreciably less energy in crop production. In addition, two new analyses have been performed that were not previously reported, which lead to two further conclusions: 4) the organic farms are drawing on the preexisting supply of P_2O_5 and K_2O in the soil, at average rates of 12 lb./A-yr and 41 lb./A-yr respectively; 5) the organic group requires 12% more labor to produce $1 worth of crop.

Chapter 3
Rodale Research

Rodale Press established a Research and Development Group in early 1974 to make possible a more intensive and systematic exploration of ideas related to organic agriculture. The R&D group has as its goal a more complete demonstration and evaluation of experimental food and both old and new methods of agriculture.

The main center for R&D's agricultural testing is the 305-acre New Organic Gardening Experimental Farm in Berks County, Pennsylvania. For over 200 years the farm was owned and operated by a German family named Siegfried. In the early 1960s the family stopped farming the farm and rented out the land. Continuous corn, with heavy applications of chemicals and limited care, took its toll on the land. Rodale Press bought the farm in early 1972, and since that time has rented 290 acres to the neighboring Mennonite farmer, Ben Brubaker, for cropping organically. A discussion of his methods appears in Chapter VI. The remainder of the farm is used for R&D's experiments and buildings.

Our research program has been directed at three primary areas. The first is an investigation of a kind of companion planting called intercropping or interplanting. Like companion planting, it offers some protection from insect and disease pests—but that isn't its main advantage. Intercropping is a way to increase the land's productivity by mixing two or more different plants in a space. Our first experiments with this technique last summer gave yields 43 percent higher overall. And other researchers around the world have discovered similar crop increases produced by older native systems of intercrop farming in Asia, Africa, and the Americas. The potential increase seems to be around 50 percent. Interplanting, like so many other facets of food production, is a simpler, better idea that needs to be rediscovered.

So are some very nutritious vegetables. And R&D's second aim is to uncover and explore these forgotten or little-known crops. We know of several that are high in protein or have properties that might be of value to a small farmer or gardener. Amaranth, with a seed grain as nutritious as wheat and leaves similar in taste to spinach but higher in protein, is our main focus now. But we have also begun to look at the winged bean, the genus *Chenopodium*, a plant called Koa Haole, holy basil, and ragi, a millet from India. They all hold terrific promise.

The third target area is capturing more of the free nitrogen in the air. Nitrogen is the soil nutrient most often in short supply. We need to utilize more of the natural, biological methods of fixation in order to increase soil fertility. The interplanting trials are part of this effort. Besides producing more crops in less space, intercropping legumes puts nitrogen in the soil for the next year's crops. Thus, R&D is examining two little-known but powerful legumes, Koa Haole and winged bean. Pleurotus Sajor-Caju is an edible mushroom which fixes nitrogen from the air. It can grow on a substrate like straw, and leave it rich in nitrogen and valuable for further composting. The association of the water fern, Azolla, with the blue green algae, Anabaena, is also potentially useful to the home gardener in a similar way.

In the summer of 1975 three acres were planted with growth trials, and the results are reported here. With one year of experience with formal agricultural testing under our belts, the 1976 effort was increased in both acreage and scope. The results of those tests and future work will be available in OGF articles, and special publications available from Rodale Press.

1975 AGRONOMIC EXPERIMENTS

Observations and trials undertaken over the summer of 1975 at the New Farm centered around the theme of greater security on the land. Major emphasis was given to evaluating protein-rich plants and developing organic methods for supplying nitrogen to crops. Many of the experiments represent the first stage of long-range projects.

In June of 1973, communications with Dr. John Robson, an M.D. directing the Human Nutrition Program at the University of Michigan, convinced Rodale Press that there were half-wild plants under limited cultivation in less developed countries that could become valuable new crops for American gardeners. A primary genus selected for in-depth agronomic and nutritional evaluation was *Amaranthus*. This many-specied leaf and grain crop is found in China, India, Afghanistan, Africa, and North, Central, and South America. In south central Mexico, Amaranth has been selectively bred for large seed yields. A white-seeded species was a well-established "back-up" grain crop there at the time of the Spanish Conquest. Although suppressed by the Spanish because of its association with Aztec religious rites, *Amaranthus hypochondriacus* has continued to be grown on a small scale in Mexico.

Data in the literature on protein quantity and quality suggested that this almost-forgotten grain crop could become a valuable addition to American gardens. Moreover, there was good reason to believe that the Mexican variety could be successfully cultivated in the United States. Closely related, black-seeded Amaranths, known as "Pig Weed," thrive in our North American temperate climate wherever there is tilled soil. We had hoped the hardy, six-feet tall Mexican plants might also produce substantial seed and green yields.

Seed obtained from Mexico was planted at the farm in June of 1974. The first step was to find out whether the Amaranth would overcome a different light/dark cycle and mature to set seed in Pennsylvania. And yes, despite a late planting date and an early frost, this observation gave promising results:

The Amaranth grew to six feet or more and set seed, although the yield was much lower than in Mexico. This was an exciting preliminary indication that Amaranth might be widely adaptable without a special breeding program.

Then more precise nutritional information obtained in late 1974 and early 1975 strengthened R&D's conviction of the value of this "new" plant. Lab reports on the *Amaranthus hypochondriacus* confirmed the reports of the grain's high protein content—15.3 percent, compared to 13.3 percent in whole wheat flour made from hard wheats.

More significantly, the quality of that protein, as revealed in its amounts of the eight essential amino acids, was unusual for a grain. Compared to egg protein, which is assumed to have the best (100) amino acid pattern for human physiological needs, whole grain Amaranth showed a protein score of 75 (VS 56.9 for whole wheat, 72.2 for cow's milk, 68 for soybeans, and 69 for a maize, beans, and groundnut mixture that is now used in underdeveloped countries as a complete protein source).

As a green vegetable the Amaranth also received a high rating. Researchers looking for plants which yield large amounts of leaf protein for extraction and animal feeds have put Amaranth in the highest category. Not only does it produce prodigious quantities of leaf matter and protein, but its protein is relatively easy to extract and digest. Although R&D isn't interested in such a processed food—still, this new scientific interest points to a quality ancient farmers have long used.

All over the tropics Amaranth has been one of the major potherbs. Lightly steamed, the upper leaves and tender shoots provide a delicious and digestible source of supplementary protein. And as greens it has an excellent supply of vitamins and minerals. Amaranth greens (called "spinach" in China) contain large amounts of iron and vitamins A and C, much like the leaves of its distant cousin, the beet.

These initial agronomic results and specific data on the high protein value of this semiwild food definitely indicated the importance of more extensive growth trials, and the first phase of a long-range research program was implemented in the spring of 1975. To build a seed stock and obtain greens for observational feeding to small livestock, block plantings of about one-fourth acre each were made of *Amaranthus hypochondriacus* and an as yet unidentified Amaranth also obtained in Mexico. Experimental plots measuring five by ten feet with a population range of 15,000 to 40,000 plants per hectare were also established and fertilized with mushroom soil compost. Selections for plant types based on color and morphological differences of flowering heads have been made from these plots.

To supplement these local, first-year results with data from other locales, several hundred *Organic Gardening and Farming*® readers were recruited to grow Amaranth as an experimental crop. By November 1975, 146 of these reader-researchers in thirty-seven states and two provinces of Canada completed the observation, returning a midseason questionnaire and a postharvest data sheet. Midseason information supplied by the participants (height measurement, rainfall over the growing period, and number of growing days) was correlated with latitude of growing locations and normal average temperature in an exercise.

Although there tended to be a relationship between height and growing days, no correlation emerged between height and latitude, annual average temperature, and rainfall. More specific data is needed for definite conclusions, but the general indication to date is that variations in microclimatic conditions do not overly affect Amaranth—a good sign confirming the versatility, wide-range adaptability, and thriftiness of the crop.

In the next field trials at the farm, the population range will be extended, and a different fertilizer regime will be used. An observation will also be done involving row plantings of seed from plants selected this year, although we do not want to sacrifice the wide genetic base of the Amaranth in seeking higher yields.

EDIBLE SOYBEANS

Dried soybeans contain approximately 40 percent protein (VS 17.9 percent for hamburger and 12.9 percent for whole eggs.) Since this protein has substantial amounts of all the essential amino acids, its quality is also unusually high for a vegetable food. For these reasons soybeans could constitute a major source of low-cost protein for Americans. At present, however, 95 percent of the soybeans grown in this country are processed into oil or animal feed, rather than eaten as a bean. And according to 1972 data, *all* dried beans and peas contribute only 1.7 percent to the national per capita protein consumption. Considering the popularity of kidney and navy beans, peas, etc., it's evident that soybeans are little used in the U.S. by the average family.

One explanation is that they are not widely available commercially either in dried or canned form. This reflects a lack of consumer demand, apparently related to the fact that dried soybeans have a much firmer texture than other beans and also must be soaked and cooked longer. Both these obstacles to increased use in the U.S. can be overcome through the availability of larger, more flavorful vegetable-type soybeans, which are more suited to being grown at home and consumed as a green or fresh-frozen vegetable. At present, however, only a few varieties of such soybeans are available to small farmers and gardeners. And these tend to be low-yielding and have poor resistance to shattering.

To encourage the development of improved types, R&D studied the yields and shatter resistance of nineteen vegetable soybeans. Varieties included in this trial were IMPERIAL, HARK, PROVAR, KANRICH, BANSEI, VERDE, KIM, FUJI, FUNK'S DELICIOUS, HIGAN, PROTANA, RA, HOKKAIDO, DISOY, WILLOMI, JOGUN, MAGNA, FISKEBY 5, and MANDELLI.

The nineteen varieties were planted 30 June 1975 with three feet between rows and one foot between experimental points (two plants per point). Two plots were established for each variety of soybean, with one plot fertilized with rock phosphate at the rate of a hundred-weight (112 pounds) per acre and the other plot not fertilized.

The varieties were observed to determine days to maturation as a fresh vegetable. Then the beans were allowed to dry in the field, and harvested by hand. In late November, the experimental crops were sorted, counted, and weighed to arrive at the number of single and double-seeded pods, sterile pods, shattered pods, and the average dry weight of beans per plant for each variety.

R&D observed no indication of any significant difference in yield between fertilized and unfertilized plots of the same variety. Therefore, in Table 3-1 highlighting preliminary data results, both fertilized and unfertilized experimental plants of the same variety have been grouped together in determining average yield per plant, etc.

In future work with edible soybeans, R&D will attempt to examine more closely the impact of phosphorus on yield. Other factors influencing nodulation—such as the trace elements molybdenum and boron—will also be evaluated.

TABLE 3-1
1975 EDIBLE SOYBEAN TRIAL

Soybean Variety	Avg. Wt. in grams Dried beans per plant	Avg. no. shattered pods per plant	Days to maturity as fresh veg.
Imperial	23.375	.406	78
Kanrich	24.959	.225	82
Funk's Delicious	21.060	.333	82
Jogun	21.828	.625	72
Ra	24.308	.167	81
Provar	20.345	.441	70
Magna	21.257	.424	65
Verde	23.606	1.24	80
Protana	30.585	.25	70
Willomi	23.585	.25	70
Hark	18.600	.971	70
Hokkaido	28.077	.621	80
Higan	42.225	.433	80
Bansei	20.859	1.31	76
Fuji	31.937	2.22	82
Mandelli	28.210	1.29	77
Kim	31.602	.444	82
Disoy	21.59	1.16	80
Fiskeby 5	7.329	.011	45

HIGH-LYSINE CORN AND EDIBLE SOYBEANS

Because soybeans are nitrogen-fixing legumes, some researchers have suggested they might provide nitrogen to corn when interplanted with that crop. High in protein, they can also give a boost to the total

food production of an interplanted area, thus
maximizing the effects of the intensive land use involved in intercropping. Interplanting vegetable
soybeans with high-lysine corn seems a particularly
good idea nutritionally, since the two crops can be
harvested at the same time. The substantial usable
protein they contain separately is increased significantly by eating them in combination—fresh, frozen,
or canned.

To determine and compare the yields of high-
lysine corn and soybeans, planted separately and
intercropped, R&D undertook a trial involving two
kinds of high-lysine corn and two soybeans (Kim
and Kanrich). The trial consisted of eight treatments
replicated three times.

Four of these treatments featured each of the corns
and soybeans alone, with the other four consisting of
one kind of corn interplanted with one kind of soybeans. The planting rate was two seeds per point
with thinning to one plant several weeks after germination, and missing points supplied.

Corn planted alone was put down in rows three
feet apart with one foot between each point within
rows. Soybean varieties planted alone were seeded
in rows one foot apart, with one foot between experimental points. For the treatments in which high-
lysine corn and soybeans were interplanted, two
rows of soybeans were planted between each row of
corn, with one foot between each row and one foot
between each point within rows.

The initial analysis was done on ten sample plants
of corn and beans from each plot. Interplanting had
essentially no effect on the yield of corn hybrid 1. As
shown on Table 3-2, the yield of corn hybrid 2 was
decreased significantly by one of the bean varieties.
With such a slight indication of reduction present in
only one of the four combination treatments, it can

TABLE 3-2
CORN-BEAN INTERPLANTING

Per 10 Sample plants	Alone mean yield per plant (net weight grams of grain)	Interplanted with Bean 1 — mean yield per plant (net weight grams of grain)	Interplanted with Bean 1 — yield difference and statistical significance	Interplanted with Bean 2 — mean yield per plant (net weight grams of grain)	Interplanted with Bean 2 — yield difference and statistical significance
Corn 1	158.4g	143.2g	- 15.2g (not significant)	159.7	+ 1.3g (not significant)
Corn 2	190.4g	180.2g	- 10.2g (not significant)	170.6	- 19.8g (significant)
		with Corn 1		with Corn 2	
Bean 1	44.08g	34.59g	- 9.49 (significant)	31.35g	- 12.73 (significant)
Bean 2	46.03g	31.50g	- 15.53 (significant)	38.27	- 7.76 (significant)

generally be said that interplanting had almost no effect on corn yields on a statistical basis.

In all four treatments, soybean yields were reduced significantly in the interplanted plots. For both varieties this amounted to a 25 percent decrease per plant.

Analyzing the data for entire plot yields confirmed this assessment of the performance of corn and beans interplanted. Taking into account a very low plant mortality in interplanting (about 1 percent), total weights showed no yield reductions for interplanted corn. That is, the slight decrease for corn 2 shown in the sample plants disappeared. Both varieties of soybeans had yields reduced by 23 percent per plant.

These first-year results show that corn maintains its productivity despite the competition created by interplanting's greater plant density. The soybeans are reduced in productivity, likely because they get considerably less sunlight through the corn's leaf canopy. And the beans do not seem to be contributing any extra nitrogen to the corn.

In total, though, production in interplanted plots is much greater. The technique produces 100 percent of a normal corn crop. Compared to a single cropped patch of soybeans, interplanting with corn leaves room for 43 percent fewer plants, each producing 23 percent less beans. So an interplanted plot also yields 44 percent of a full bean crop—or total production of 144 percent from a piece of ground.

"Intercropping has the potential to create tremendous yield increases. In fact, there's almost no way you can lose with intercropping, at least in theory," says R&D agronomist Alex Cunard. Studies in India have shown that nitrogen fixed by legume companions can be used by the grain companion for higher production. There is some evidence from South American research that this will work with the corn-soybean combination. To help overcome reduced sunlight, future trials will attempt to encourage N-fixation by inoculating the beans in one trial. Others will use new planting patterns and spacing, and ideal soil pH and trace mineral levels.

HIGH-LYSINE CORN, SOYBEANS, AND ALFALFA

Another interplanting experiment begun by R&D in the summer of 1975 involved not only high-lysine corn and soybeans, but alfalfa—another legume capable of fixing nitrogen and providing it in a form available to corn. The aim was to observe the performance of seven high-lysine corn hybrids with soybeans and alfalfa to determine yields of corn; yields of alfalfa; yields of corn and alfalfa interplanted; yields of soybeans; and yields of corn and soybeans interplanted.

The planting pattern in this observation was rows of corn planted three feet apart, with one foot between plants in a row; rows of Kim soybeans, planted one foot apart, with one foot between each plant in a row; and Buffalo alfalfa drilled in rows one foot apart. There were seven treatments, one for each variety of the corn.

In each treatment, three rows of corn planted alone merged into an intercropped area featuring two rows of soybeans between the corn rows. This interplanting then became a planting of soybeans alone. The remaining two rows of corn in each treatment merged into corn interplanted with two rows of alfalfa between each row. This merged into a planting of alfalfa alone. Both corn and soybeans were planted at two seeds per point, and the alfalfa was seeded in drills at the rate of four pounds of seed per acre.

During the growing season, the problem of weeding these plots was very real. It was observed that the corn as a sole crop, which could be and was weeded with a rototiller, grew better than the corn in the interplanted rows, which was more subject to weed competition until manual weeding was done. The alfalfa appeared to suffer most from the close interplanting. Although it germinated very well, weeds suppressed growth, and there was no yield at all from the interplanted alfalfa.

Data tabulation has been done using two plants taken at random from the corn area, the soybean and corn interplanted area, and the soybean area of each treatment. Calculated yields per acre were obtained by multiplying the average yield of the two test plants by the population per acre. Since a complete stand is never available over an area such as an acre, the calculated yield was multiplied by 0.8 to give an adjusted yield per acre that would be a better estimate of the performance of a variety under the conditions of the observation. The listed yields are low, but were not intended to show maximum potential crop returns, but to gauge the effect of interplanting.

The adjusted yield per acre for each variety of

TABLE 3-3
ADJUSTED CALCULATED YIELDS PER ACRE

High-Lysine Corn	Adjusted Yield of Corn as Sole Crop (Tons)	Adjusted Yield of Corn When Interplanted (Tons)	Total Adjusted Yield of Corn 1 Soybean Interplanting (Tons)
Pioneer L3369	0.456	0.168	1.480
Pioneer L3516	0.696	0.392	1.264
Pioneer L3579	0.464	0.576	2.008
Crow's HL630	0.584	0.424	1.304
Crow's HL711	0.528	0.672	1.360
Crow's HL284	0.544	0.192	1.440
Crow's HL450	0.312	0.312	1.936

high-lysine corn planted alone is given in Table 3-3, which also shows the yield for the corn when inter-planted and the total adjusted yield per acre obtained when corn and soybeans were interplanted.

In comparing yields of soybeans planted alone and those interplanted with the corn, we found that in four treatments there seemed to be an increase in soybean yields with interplanting. However, the average result was a reduction in yield. As far as corn yields were concerned, in only two treatments out of the seven was there an increase in the yield of a variety with interplanting. Generally there was a reduction in yield. However, as Table 3-3 indicates, this observation confirms that interplanting produces substantially greater total yield per acre through intensive utilization of space, even though the individual yields of the interplanted crops averaged out lower.

Future work with corn, soybean, and alfalfa inter-planting will involve planting legumes within corn rows to allow mechanical cultivation. This will reduce weed competition to test whether these legumes can increase corn yields.

VEGETABLE AND LEGUME INTERPLANTING

The third in R&D's series of interplanting experiments involved the intercropping of selected vegetables and legumes in an effort to determine whether the residual nitrogen in the root nodules sloughing off the leguminous plants would be available to the roots of companion-planted vegetables. Most literature indicates that the general effect of interplanting is an increase in yield of the non-leguminous companion, and our goal was to duplicate this effect. Specifically, the observation was undertaken to ob-

tain information on performance and yield of each
vegetable crop, performance and yield of each
legume crop, and performance and yield of each
vegetable and legume combination.

Five different vegetable-legume combinations
were chosen, and the legume served as the base crop
in four of the five pairings: lettuce-kidney beans,
kohlrabi-blackeyed peas, tomatoes-soybeans, and
beets-soybeans. In the fifth, lima beans was the base
crop interplanted between rows of carrots. The spacing between plants in the monocropped patches approximated common practice for all varieties.

Since the habits of growth certainly influence
compatibility of two plants, Dr. Cunard worked a variety of types into this experiment. He chose two
spreading beans, limas and kidneys, monocropped
in rows two feet wide, as were their companions in
the monocropped stands. The bush-type soybeans
and blackeyed peas were monocropped in rows one
foot wide, and paired with vegetables monocropped
three feet apart. This resulted in one-foot rows
between companions in all interplanted treatments.

The vegetable companions represent root, fruit,
and greens crops. All except the kohlrabi were
harvested as they came ripe and weighed at full
moisture. The kohlrabi was cut at the first frost, and
the beans were vine-dried and harvested in mid-
autumn.

The results of the observation are summarized in
Table 3-4. Though there is a wide range apparent in
the effects of interplanting, Dr. Cunard points out
some interesting information this experiment revealed. Interplanting produced little or no adverse
effect on the bean yields, except the reduction in the
lima-carrot combination. In beets-soybeans, the bean
yield per plant is substantially increased. For every
vegetable companion except lettuce, interplanting

TABLE 3-4
VEGETABLE—LEGUME INTERPLANTING

Replications		Planted Alone avg. yield per 6' x 8' plot	interplanting reduces no. of plants per plot	Interplanted avg. yield per 6' x 8' plot	avg. effect on yield per plant
Lima beans	3	1.93 lbs.	0	1.39 lbs.	28% reduction
Carrots		92 lbs.	25% (or 1 row/plot)	57 lbs.	17.4% reduction
Kohlrabi	2	25.25 lbs.	0	9 lbs.	64% reduction
Blackeyed Peas		4.84 lbs.	36.3% (or 3 row/plot)	4.57 lbs.	0 effect
Lettuce	3	1.93 lbs.	0	2.26 lbs.	17% increase
Kidney Beans		3.08 lbs.	28% (or 1 row/plot)	2.08 lbs.	5.8% reduction
Tomatoes	1	118.5 lbs.	0	71.5 lbs.	39.6% reduction
Soybeans		11.1 lbs.	51% (or 3 row/plot)	5.47 lbs.	0.2% reduction
Beets	2	21.25 lbs.	0	7.25 lbs.	66% reduction
Soybeans		10.63 lbs.	41.3% (or 3 row/plot)	8.50 lbs.	36.4% increase

reduced the crop; Dr. Cunard says that all were severely shaded by the beans. The beets and the kohlrabi were almost completely suppressed by the overgrowth. Apparently one-foot row spacing is too narrow for these crops. The lettuce, however, was harvested long before the kidney beans reached full size. In this case the shading provided by the bean plants may have been the very cause of the superior growth of the lettuce.

The legume companions chosen for this experiment may also be a factor in the performance, says Dr. Cunard. Theoretically any one of them can fix all the nitrogen it needs, but some are not very efficient. Early peas and snap beans, for example, fix very little. In this observation, none can match the soybeans (with a capacity of 100 pounds per acre a season). Next best are the blackeyed peas (around 70 pounds per acre). In future trials, different vegetable legume pairs and planting patterns will be tried in an effort to move closer to determining those combinations best suited to the intercropping technique.

POTATO MULCHING

Mulching is an efficient technique that has long interested organic growers. It promises a multitude of benefits for nearly any plant, affording water conservation, soil temperature moderation, and organic fertilization. Where potatoes are concerned, the practice has proved especially attractive, because the tubers consistently form at the soil surface, just under the mulch. Soil-free, they require no digging, and are within easy reach for a few early meals. Some gardeners even claim greater yields. Others, however, have experienced total crop failures with mulching.

Potato-mulching trials were established at the farm to test responses to three different types of mulch — grass clippings, leaves, and straw.

This trial was designed to find what difference mulching makes in yield, and beyond that, what kinds of yield increases might be attributed to using one material rather than another. So three readily available mulches (leaves, grass clippings, and straw) were compared to weeded and unweeded unmulched controls.

All five treatments were tested on two varieties of potatoes, the brown Katahdin and the red Chieftan. Trial plots were two feet by five feet with three rows per plot. Altogether, there were fifty-seven plants per plot. Each distinct treatment was replicated three times.

In the second week of May, the soil was tilled to a depth of six inches; no fertilizer was used. Each

Organic Farming

piece of seed potato, cut with at least two eyes, was planted one foot apart in a very shallow drill, with enough soil to just cover the eyes. (Surface planting seems to encourage surface tuber formation, but rodents have been known to destroy surface-planted seed, thus, R&D's planting technique was a compromise.)

The respective mulches were used in the forms available to gardeners. Raked the previous fall, the leaves had been stored in the open and were partially decayed. Likewise, the straw had been left out all winter. Grass clippings, however, were fresh-cut and lawn-dried. Each mulch was applied six inches deep. The unmulched weeded treatment was weeded twice. None of the mulched plots needed weeding.

Before harvest the researchers noted greater vigor, evidenced by color and size, in the vines under leaf and grass mulches. The leaf mulch kept the plants green for a week longer than all the other treatments except grass clippings, and the grass clippings maintained the plants a week longer than leaf mulch.

All the crops were harvested during the second week in August after the last tops had browned. As hoped, the tubers all formed near the surface and were easily harvested with a trowel. The average yield per plot under grass clippings was 47 pounds of potatoes; under leaves, 44 pounds, and under straw, 37.5 pounds. The weeded control plots averaged 27.8 pounds while the unweeded produced only 14.7 pounds. Table 3-5 summarizes these findings.

This trial verified that mulching potatoes will produce substantially larger crops than traditional hoeing and hilling. All three mulches increased yields significantly, although R&D's work so far suggests grass and leaves are better than straw. Yields under mulch can be increased even more dramatically in nonexperimental situations because this method also permits a much denser planting pattern.

TABLE 3-5
POTATO MULCHING RESULTS

Treatments	I Ch.	I K.	II Ch.	II K.	III Ch.	III K.	Tot.	Avg.
Grass Clippings	58	55	41	42	40	46	282	47.0
Decayed Leaves	42	35	49	50	42	46	264	44.0
Straw	35	29	33	41	50	37	225	37.5
Unmulched Weeded	27	32	28	26	29	25	167	27.8
Unmulched Unweeded	8	14	14	17	11	24	88	14.7
	170	165	165	176	172	178	1,026	

RAGI

Because of climate, farmers in the southern U. S. are limited in the number of small-grain crops they can grow successfully in the summer. R&D is hopeful that experiments with ragi (Eleusine coracana) will result in a versatile new small grain for such farmers.

Grown in Africa and India, ragi's primary cultural requirement is heat (a mean temperature of 75° F.) which it must have through its entire growing period. Growth is best with medium rainfall. But, like sorghum, it stands drought, having the ability to halt growth and await rain. Ragi does well under irrigation, and is very tolerant of the salinity that often builds up in irrigated soil. Its adaptability to high altitudes and its variable (three to five months) maturing time further extend the useful range of this grain.

Also known as finger millet, ragi has long been a staple in North Africa and south India. In India, its reported storage life of fifty years has earned ragi fame as insurance against famine. This reputation also rests on its high biological value: some strains are as nutritious as wheat. Although not closely related to sorghum or other grains called millet, on the average ragi has a similar nutritive value: 7.1 percent protein, 1.3 percent oil, and 76.3 percent carbohydrate. One white variety, however, offers 14 percent protein. Ragi is excellent for making malt which is often made into beer or used dry. More commonly the grain is served in flat breads or porridges.

Even though the Pennsylvania summer might be too short for this tropical plant, R&D was eager to try out such a potentially valuable food. Researchers obtained seeds of four varieties of ragi (EC4849; EC4840 CO10; PR202; CO7) for observational experimentation. The seed's late arrival further shortened the growing season, delaying the planting until late June. Still, the results were promising.

Germination was good, and healthy stands (two rows, forty-five feet each) of all four types were established. One ragi was low and spreading in habit, while the other three were upright. Since none were close to setting seed by the end of August, we built a wooden frame around a row of one of the upright varieties, and covered it with plastic. This insured higher temperatures for the plants within. Most of these plants did form seed heads, which were harvested at the end of September. Although the uncovered rows formed flowering heads, none set seed. Future trials will concentrate on earlier plantings.

FISH-FARMING PROJECT:

The fish-farming experiment at the New Farm was inspired by techniques for raising carp first developed over 4,000 years ago in China. In these ancient systems, farmers cultured small ponds so that they were nearly self-regenerating, like miniature ecosystems. Animal manure was added to stimulate the growth of microscopic aquatic life, which, along with kitchen wastes, served as fish food. In turn, the ponds yielded nutrient-rich irrigation for the fields and, of course, fresh fish for the table.

The nutritional excellence of fish stirred R&D's interest just as much as the ancient technique's efficiency. As rich in protein as beef, fish is more easily digested and contains only about one-fourth the fat of lean beef. Moreover, trimming losses are lower: 40 percent for fish compared to 60 percent for beef. All in all, the researchers felt sure that developing a backyard fish-farming system would create a valuable new option in home food production.

For the first phase of the experiment, we decided to develop a backyard aquaculture system which could supply a family of four for a year.

According to yearly *per capita* consumption figures, such a pond would have to grow at least 80 pounds of fish in a six-and-one-half-month season. Secondary objectives were to explore techniques which require a minimum of equipment, time, and money. Lastly, the researchers wanted to see how much irrigation with nutrient-rich pond water could increase yields in a vegetable garden.

After consulting with aquatic biologists Dr. James Avault and Dr. Homer Buck, the staff constructed six test ponds, each twelve feet in diameter and three feet deep (339 cubic feet). This size was calculated to provide about four cubic feet for one pound of harvested fish, more than required with ideal aeration. The consultants recommended this wide margin since we wanted to minimize equipment for aerating.

Aware that water quality (*i.e.*, oxygen, temperature, pH, and ammonia levels) would be the critical factor in determining the growth of the fish, the test was designed for algae to produce oxygen. The plants would also use up ammonia excreted by the fish to keep it from increasing to toxic levels. To encourage a rich growth of algae, every pond was fertilized with 1.6 pounds of manure. Each pond also had a small bubbling aerator for oxygenation.

On 22 May, each of three ponds was stocked with twenty catfish fingerlings (.15 lbs. each) and ten buffalo fish (.04 lbs. each). The other three each received ten mirror, or Israeli, carp (.3 lbs. each), ten buffalo fish, and sixty-one crawfish (.02 lbs. each).

Being evaluated elsewhere for commercial cultivation, these warm water species were chosen for palatability, hardiness, and efficient feed conversion. By combining species in a pond, the staff anticipated a more efficient use of the available food.

In addition to foraging on plants and animals growing in the pond, the fish were given as much of two prepared feeds as they would take. To see if backyard fish farmers might save by feeding table scraps, the staff ground up wastes from the Fitness House kitchen according to a rule-of-thumb formula. This feed was made up as follows: 50 percent from high protein sources (meat, cheese, etc.), 25 percent from high carbohydrate scraps (bread, rice) and 25 percent from vegetable wastes (vitamin-rich "greens"). The kitchen-derived feed went to the carp ponds, while the three catfish ponds got a commercial floating trout chow (40 percent).

The ponds were harvested on 21 October. Average

weights per fish were .8 pounds for catfish, 1.5 pounds for carp, and .6 pounds for buffalo. Nearly all the crayfish had disappeared: the nine remaining averaged .04 pound each. Although the total yields were much lower than had been aimed at, the R&D staff was extremely heartened by these table-sized fish. Several fisheries experts had said that it was simply impossible to produce fish for food in such small ponds. Yet these catfish and carp were over half the size the staff had projected under the best possible conditions, and a late start (a month and a half or 25 percent of the ideal) had reduced the outcome before the experiment had begun.

The rates of gain represented by these figures compared well with those obtained in other trials in the literature. As this implies, both feeds nourished the fish well. The catfish and the buffalo on the commercial rations converted every 2.2 pounds of feed to 1 pound of meat, coming close to the 2:1 ratio well established by previous research and commercial performance. Preliminary observations indicate that the table scrap feed was equally effective, producing the same weight gains per pound of feed in a dry matter comparison.

As expected, the garden irrigated from the ponds produced more than its counterpart watered directly from the spring. All pond-watered vegetables grew better; total yield averaged about 30 percent greater.

On the basis of these first-season results, the R&D staff is confident that fish can be grown successfully in backyard-sized pools. However, they have learned that aquaculture is not simple on a small scale and with limited technology. Just obtaining good stock was a problem. Not only were the buffalo fingerlings shipped ten times smaller than ordered: they also carried anchor worm disease from the hatchery. This parasite destroyed 90 percent of the buffalo within the first two weeks.

Oxygen levels in the test ponds also fluctuated, with serious results. For example, in late August an aerator failure in one of the carp ponds suffocated the fish within twenty-four hours. All the ponds, especially the three with carp, often had marginal oxygen levels in the latter half of the season.

At the root of the oxygen problem was severe and constant leaking. The catfish ponds lost 25 percent of their water daily, and the carp ponds lost 50 percent. Adding the necessary quantities of 55° spring water slowed R&D's warm water fish down and greatly disturbed algal growth. The three carp ponds supported visibly less algae than the catfish ponds. Consequently their oxygen levels and fish growth rates were also substantially lower.

Future trials will ensure a stable ecology in the ponds by sealing them. Plans still call for algae to supply most of the oxygen and they will pursue ways to maintain the optimum algal population. If they are successful, aerators will be needed only as a supplement in hot weather. When the staff obtains the expected oxygen levels, the stocking rate will be doubled, making possible a yield of 160 pounds per pond.

Chapter 4

Soil Organic Matter

Microbes and What Happens in the Soil

Each spoonful of arable soil contains billions of living microscopic organisms. Multiply this by the number of spoonfuls of soil in an acre and you have figures that are astronomical. The living bacteria alone in an acre of soil of average fertility would weigh as much as a medium-sized dairy cow.

This seething mass of microorganisms constitutes a crop of three to five tons per acre-foot of soil that the farmer sustains beneath the surface in addition to the crop that he grows above the ground. If the crop of microorganisms beneath the surface does not have adequate food, the crop above ground may suffer from competition for mineral nutrients and be more susceptible to disease.

"Microorganisms eat at the first table." They are in contact with almost every particle of soil. Plant roots are not. Without microorganic life, soil—the dynamic perpetual system that sustains terrestrial life—would become an inert mass incapable of providing food. Microorganisms decompose organic material and release elements and organic food for repeated use.

The soil and its environment provide a location for the individual microorganisms to grow and multiply. The soil's solid phase is composed of rocks, stones, sand, silt, clay, and plant and animal residues in various stages of decomposition. Generally, the finer particles are arranged or grouped into clusters called aggregates.

Between the soil particles are air spaces containing the atmosphere of the soil. The composition of soil gases is generally the same as for air gases— about 80 percent nitrogen and 20 percent oxygen. Usually, the carbon dioxide concentration is higher in soil than in air. Other gases, such as methane and hydrogen gas, may be present under anaerobic conditions which may prevail under high moisture and poor drainage conditions.

As plants grow, roots spread through the soil. The presence of root exudates and soluble salts around living roots creates a favorable medium for microbial activity. Microorganisms may be ten times more concentrated around and adjacent to roots than away from them. Also, the kinds of organisms near the roots may be quite different from those only a few centimeters away.

Although the mineral part of the soil is important, most of the microorganisms in the soil live on dead plant or animal material. However, a few flourish on living plants and animals. If they harm the plant or animal, they are called pathogens. The great majority of microorganisms in the soil perform one or more beneficial functions.

Living microorganisms in the soil include bacteria, actinomycetes, fungi, viruses, protozoa, yeasts, algae, nematodes, and other organisms.

Viruses

These forms of life are so small they can be seen only with the electron microscope that magnifies 10,000 times or more. Some viruses have been purified and precipitated as chemical compounds. Viruses have all the properties of living organisms, are able to reproduce themselves and grow, and require food for carrying on life processes. They obtain their food from living things—bacteria, higher plants, or animals. Viruses that live on bacteria may cause the bacterial cell to disintegrate soon after penetration. Many soil bacteria have viruses that live on them. These viruses may kill the bacteria they live on and other neighboring pathogenic microorganisms.

Bacteria and Actinomycetes

Bacteria are too small to be seen with the naked eye but can be seen through an ordinary light microscope. It takes about 25,000 bacteria lined up end to end to measure an inch. These are of various shapes and sizes, such as rods, spheres, and spiral forms. Some bacteria form spores which are tolerant to adverse environmental conditions, such as lack of moisture or unfavorable soil temperature.

Bacteria are very simple plants that reproduce by each cell splitting into two new cells. Bacteria perform many beneficial functions in the soil, such as fixing atmospheric nitrogen, decomposing crop and chemical residues, and stabilizing soil structure.

The actinomycetes are more complicated than bacteria, and generally have a branched cell structure. About 10 to 30 percent of the micoorganisms in the soil are actinomycetes. Actinomycetes decompose crop residues and increase the availability of mineral nutrients to crops.

The characteristic odor of newly plowed soil in the spring is due to the activities of actinomycetes. The actinomycetes produce a large variety of colored pigments.

Fungi

This group of microorganisms varies widely in form and shape and exists in all soils. Fungi are multicellular and larger than bacteria. Structure of the fungus cell can be seen in much greater detail with a low-power microscope than can that of the bacterial cell.

Fungi perform such beneficial functions in the soil as decomposing crop residues, increasing soil aggregation, and increasing the availability of plant nutrients. Fungi grow best in an aerated soil and many of them grow readily in an acid soil.

Algae

These are microscopic plants that form chlorophyll in the presence of sunlight. They can often be seen as a green scum on ponds or on the soil surface during wet periods. In the Great Plains, they can be seen frequently on the soil surface of wheat fields in the late spring. In the presence of sunlight they change carbon dioxide from the air into organic matter. Algae also live beneath the soil surface, where they feed on organic matter as do other microorganisms.

Yeasts

These single-celled microorganisms are similar to bacteria except they are larger and their structure is more highly developed. Yeasts make up only a small percentage of the total number of microorganisms in the soil. These microorganisms break down organic matter of the soil and release mineral nutrients.

Protozoa

There are many kinds of this microscopic form of life in the soil. They are almost large enough to be seen without magnification and many are mobile. Some may become inactive by changing into a cyst form that is resistant to adverse environmental conditions. When conditions become favorable, the protozoa may reverse its form and resume normal growth. They are important in the overall balance of the soil microbial population. Large quantities of bacteria are consumed by protozoa.

Nematodes

Nematodes are spindle- or cylinder-shaped nonsegmented worms about one-tenth of an inch long and may number thousands in a pound of soil. Some live on dead plant and animal material, but others infest living microorganisms and higher plants and are capable of causing root disease of plants and interfering with normal growth.

Other Organisms

There are many other living organisms in the soil which are visible to the naked eye, such as earthworms, ants, crayfish, moles, gophers, and rodents. These organisms play an important role in mixing and aerating the soil while devouring and decomposing plant and animal residues.

The soil microorganism population changes drastically with time, which affects plant nutrient availability. The amount of nutrients available for plant growth depends to a large degree on the activity of these and other forms of life in the soil.

Close examination of decomposing organic matter shows it to be teeming with soil microorganisms, indicating that organic matter provides an energy source for the active microorganisms. Because organic matter content is highest in the soil surface, soil microorganisms are also most numerous in this zone.

It is this combination of decomposable material, adequate water and oxygen supplies, and a dynamic microbial population which helps maintain the soil at a level of fertility capable of supporting higher plant growth.

Since soil microorganisms produce and secrete substances called enzymes that break down most materials that are added to or are present in the soil, the soil is virtually one big digestive system.

For example, if a material such as wheat straw, which is largely cellulose, is added to the soil under favorable moisture and temperature conditions, it is rapidly broken down to sugars. These sugars are readily used by another group of microorganisms which either decompose sugar into carbon dioxide and water or convert it into other chemical compounds.

Some of the straw may be changed into dark-colored substances that persist in the soil for a longer period of time. This is usually designated as humus. Some of the most resistant parts of the straw, such as lignin, may remain undecomposed for a long time. This adds to the humus.

Soil microorganisms have diversified growth requirements. Some organisms, such as nitrifying bacteria, can use ammonia and nitrates for sources of energy and can synthesize new protoplasm in a simple mineral medium. Many soil microorganisms, however, require certain organic compounds often found in crop residues as their energy source. These compounds, in addition to inorganic plant nutrients, are used as building materials for creating new cells.

The organic substances required from crop residues by the different soil microorganisms vary considerably. Some organisms use cellulose. Others use sugar only after it is converted from cellulose. Thus, it is necessary for certain organisms to prepare the food for other organisms.

Higher plants do not depend on soil microorganisms or crop residues as a source of organic substance. They synthesize their own organic compounds from carbon dioxide and water in the presence of sunlight and mineral nutrients.

Plants need about the same mineral nutrients as do soil microorganisms. If the soil supply of nitrogen, boron, phosphorus, potassium, copper, sulfur, magnesium, iron, manganese, zinc, molybdenum, or calcium is limited, the higher plant often will not grow normally. Microorganisms are aggressive competitors; consequently, they may fill their needs before the plant is capable of extracting its requirements.

SOIL ENVIRONMENT

Among important soil environment factors that influence the number and activity of soil microorganisms are temperature, moisture, aeration, and acidity or alkalinity. Some of these factors can be controlled by the farmer; others appear to be beyond his control at the present time.

During the warm part of the year, microorganisms may be very active if other soil conditions are favorable. When the soil temperature drops below 60°F., microorganisms slow down considerably. At soil temperatures of 40 to 50°F., some activity still will take place but, when soils freeze, the activity of microorganisms virtually stops. The soil is then in a state of preservation.

For example, the nitrifying organisms are most active at temperatures between 70 and 100°F. Below 70°F., they become progressively less active. At 50°F. or below, they produce very little available nitrogen and their activity practically ceases at slightly above freezing.

Since little or no available nitrogen is produced in cool weather, some crops may evidence a nitrogen deficiency in early spring, especially when rainfall is above average. Management practices that lower soil temperature (for example, heavy applications of straw mulch) will decrease the activities of the nitrifying organisms. Light applications of straw mulch have little or no adverse effect on soil temperature or nitrate production.

Water is essential for growth of microscopic plants and animals. When the soil moisture content drops to a level comparable to, or lower than, the wilting point of higher plants, activity of most microorganisms virtually ceases and the soil organic matter is again temporarily preserved.

Because the liberation of nutrients from soil organic matter and soil colloids is greatly decreased in a dry soil, the availability of nutrients to microorganisms is severely limited. As a result, many microorganisms die; others change to the more resistant form of spore or cyst. Some organisms, such as fungi, can grow on plant residues at a low moisture content. Also, a very high humidity may supply sufficient moisture for fungi to grow in a dry soil.

When a soil that has been dried is rewet, the microorganisms will revert from a resting stage to actively growing cells. This greatly increases microbial decomposition of soil organic matter, with resultant increased liberation of plant nutrients. Thus, following a rain, crops may often show rapid growth which is due to these additional available nutrients as well as to the moisture.

The optimum moisture content for maximum nitrification (conversion of ammonia to nitrates) is about 60 percent of the water-holding capacity of the soil. Nitrification practically ceases at low moisture content. However, if the soil is too wet, as during prolonged periods of wet weather or over-irrigation, nitrates may disappear from the soil by denitrification (biological reduction of nitrate or nitrite to gaseous nitrogen (molecular nitrogen or the oxides of nitrogen). The process results in the escape of nitrogen into the air and, hence, is undesirable in agricultural soils.) Since nitrates are water-soluble, they may also leach to depths not reached by plant roots.

All soil microorganisms require oxygen in some form—either free as exists in the atmosphere, or combined with some substance as occurs in plant or animal materials. If the soil is well granulated and not excessively wet, it generally will supply sufficient oxygen to meet the requirements of most crop plants and microorganisms.

Nitrification takes place only in the presence of air. Although tillage of the soil is aimed at keeping the crop weed-free, good aeration of the soil is also accomplished, and this promotes the activity of the nitrifying bacteria.

Acidity greatly decreases the activities of microorganisms. Fungi generally tolerate higher acidity than do bacteria or actinomycetes. Calcium may neutralize the acidity of soil. In addition, legume bacteria and azotobacter require calcium for growth. Calcium is also necessary for the activity of the nitrifying bacteria.

Many farmers have observed that liming a soil, which supplies calcium, may increase crop yields of nonlegumes. This may be the result of increased nitrification caused by liming, thus making more nitrogen available. However, lime may cause more rapid decomposition of the soil organic matter and, in the long run, unless more crop residues and adequate fertilizer are returned to the soil, cause a decline in soil fertility.

NITRIFYING BACTERIA

Many organisms are involved in the decomposition of crop residues and soil organic matter before the nitrifying bacteria can go to work. Several groups of microorganisms are involved in breaking down the complex organic nitrogenous substances in the soil to simpler nitrogen compounds. These are then broken down into ammonia nitrogen. In changing ammonia to nitrate nitrogen, a process called nitrification, two groups of microorganisms participate. The nitrite bacteria change ammonia to nitrites. The nitrate bacteria change nitrites to nitrates.

Nitrifying organisms are present in all cultivated soil, but their numbers may vary from a few up to many thousands per gram of soil. Although they usually are confined to the surface eighteen inches of soil, sometimes they are present to a depth of six feet in well-aerated soils.

Fertilizers, particularly lime, phosphorus, and potassium, stimulate the activity of the nitrifying bacteria. The response of a crop to a single-element fertilizer may often be enhanced by the fertilizer's stimulus of the nitrifying bacteria. A growing crop may be greener after lime application, not necessarily because of the effect of calcium on the crop, but because the lime has decreased the acidity and thereby increased the nitrifying bacteria's ability to produce nitrates.

In the process of breaking down or decomposing straw, the mineral constituents of the straw (nitrogen, calcium, magnesium, phosphorus, potassium, etc.) may be left behind in an available form to be used by the plant. This mineralization of plant and animal residues and of soil organic matter should increase the supply of available nutrients.

However, plant material, such as wheat straw, may contain too little nitrogen to be a balanced diet for soil microorganisms active in the decomposition process. In this case, certain microorganisms consume the carbon in the straw and the available nitrogen in the soil to obtain a balanced diet. Until their activities slow down or cease, there may be little or no available nitrogen remaining for the plants. This is called biological "tie-up" of available nitrogen and may create or intensify the deficiency of available soil nitrogen.

If much straw is added to the soil, nitrates may largely disappear for a time. Any plant material with a low proportion of nitrogen relative to carbon (high carbon-nitrogen ratio), when added to a soil low in available nitrogen, may cause a nitrogen deficiency. This does not mean that the growth of the nitrifying bacteria is depressed but that the available nitrogen they produce is used by the microorganisms in the decomposition process. After four to six months, the nitrates thus tied up by microorganisms decomposing the plant material will again be available for

143

plant use.

When a plant material, such as a legume (alfalfa), is added to the soil, available nitrogen is released upon decomposition. This is because a legume has a low carbon-to-nitrogen ratio. This means that there is a surplus of nitrogen in the residues as a diet for the microorganisms.

Generally, a plant residue should contain at least 1.3 percent nitrogen in order not to create a nitrogen deficiency during its decomposition. An excess of 1.3 percent of nitrogen in the plant material will lead to an increase of available nitrogen in the soil. The kind of residues affects the ease of decomposition, but this is also subject to some degree to the nitrogen available either in the plant, from the soil, or added as fertilizer.

Growth activities of nitrifying bacteria and higher green plants are similar. During the winter there is little or no activity of the nitrifying bacteria and higher plants are dormant.

As spring comes, the nitrifying bacteria start to produce available nitrogen. Green plants also commence slow growth in the spring. As the season warms up, bacterial and plant growth speed up.

As summer approaches, the activity of the nitrifying bacteria slips into high gear. This coincides with the period of maximum nutritional needs for the development of the green plant, such as corn.

When the surface soil dries, the activity of the nitrifying bacteria slows down. At this time, plant growth also slows down because of limited available moisture.

Most available nitrogen is bacterially produced in the surface foot of soil. Since nitrates are soluble, they are carried down with the percolating water. Roots of a growing crop absorb much of the available nitrogen and transform it into plant tissue; if excess water moves through the soil, however, there may be losses of soluble nitrogen and other plant nutrients.

As a soil dries, subsurface moisture rises by capillary action, bringing with it water-soluble salts. Thus, some of the nitrates that have been carried downward by percolation are moved upward where they can be more readily utilized by plant roots.

Use of plant residues that contain a large quantity of nitrogen relative to the carbon content will insure production of nitrates surplus to the needs of the soil microorganisms. For example, legume residues are excellent sources of available nitrogen.

Placement of residues also influences the activities of the nitrifying organisms. When residues are left on the surface as a mulch, the nitrifying bacteria are usually slightly less active than when the residues are plowed under. This is true with both legume and nonlegume plant material.

Straw plowed under usually results in a temporary depression of nitrates. If the residues are legumes, plowing them under increases the available nitrogen supply of the soil soon after plowing. These alternatives or combinations of them allow the farmer some control over the nitrification. The method or combination of methods selected depends upon the soil and probable rainfall.

The use of stubble mulching, together with the use of nitrogen-fixing legumes, may permit the production of legumes in dry areas that could not grow them before because of extreme moisture extraction by the legume. The low nitrate production with stubble mulching, plus the additional moisture that occasionally may be stored with the stubble-mulch system, may often increase crop yields in drier areas and, at the same time, afford erosion control.

During fallowing of residue-covered land, there is usually enough moisture to stimulate abundant nitrate production; thus, accumulation of nitrate usually is greater under fallow conditions than during cropping. In the Great Plains area, as much as 200 pounds of available nitrogen per acre may be produced in a single fallow season, as compared with a production of 20 to 100 pounds of available nitrogen on land in row crops.

AVAILABILITY OF NUTRIENTS OTHER THAN NITROGEN

Conditions favorable for microbial activity usually coincide with conditions favorable for plant growth. A large part of the nutrients for plants is stored in organic form and, thus, is not readily available to plants. Under favorable conditions, microorganisms change approximately 1 percent of the total nutrients in the organic matter to an available form.

For every ton of straw or plant residues decomposed by microorganisms, 1½ tons of carbon dioxide are produced which may form carbonic acid. This acid has a solvent action on the soil minerals that may change inorganic nutrients to an available form for plant use. When nitrogen and sulfur are oxidized in the soil, nitric and sulfuric acid are produced. These acids in localized areas undoubtedly have a dissolving action on the soil minerals.

Bacteria, actinomycetes, fungi, protozoa, and algae constitute the kinds of microorganisms in the soil. Their numbers range from a few million to several billion per gram of soil.

Soil microorganisms are involved in many beneficial activities in the soil. These are decomposition of crop residues, mineralization of soil organic matter, synthesis of soil organic matter, nitrification, fixation of nitrogen, immobilization of mineral nutrients, and formation of organic substances which may be both stimulative and toxic to plant growth, depending upon concentration. Organic substances formed by microorganisms may influence soil structure stabilization.

Many of the cropping and tillage practices that a farmer uses are effective in crop production because of their influence on microbial activity. For example, when the soil is tilled, aeration may be improved, and aeration is favorable for the growth of the nitrogen-, sulfur-, and iron-oxidizing organisms. When the soil environment lacks oxygen, it is unfavorable as an environment for many plants. Legumes are inoculated, planted, and turned back into the soil to increase available nitrogen for the following crop.

Every practice or management system influences microbial activity which, in turn, influences the decomposition of plant residues, the availability of nutrients, and the soil structure. These all influence crop growth, and the growth of the crops determines the soil cover and the erosion protection afforded.

The effects of microorganisms on herbicides are much more apparent than the converse; there can be little doubt that the application of a herbicide to a soil, with all of its complex and dynamic systems, alters the balance between the microorganisms present.

Fertility of the soil is governed to a great extent by the balance between the various types of microorganisms whose actions play a major role in the carbon, nitrogen, and other mineral cycles.

T. M. McCalla
From a talk at the "Organic Residues and By-Products In Crop and Animal Production Workshop," held by the University of Nebraska, Dec. 1975.

What Balanced Fertility Really Means

The premise that profitable, higher quality crops can be obtained by using less rather than more manufactured, soluble chemical fertilizer, or perhaps not using any at all, goes against the grain of everything the conventional farmer has been taught to believe. Corner him in a lush organic garden where he can't deny what his eyes see and he will only laugh and say "Sure, you can do that on a little-bitty-old garden patch, but you can't do it on 5,000 acres."

Maybe not, but there's a growing group of independent and idealistic commercial farmers who think you can do it on 500 acres or less and make enough net profit so you don't have to farm 5,000. They call themselves natural farmers, or sometimes eco-farmers, or occasionally organic farmers. I will stick with the term "eco-farming" because these fellows are certainly not natural in Rousseau's sense of the word, nor are most of them completely organic in the Rodale definition of the term. They all believe, however, with organicists, that straight chemical farming is, in the long run, too costly both to their pocketbook and to the environment.

If recent studies, particularly those done by Washington University, are any indication, that belief is justified. If so, it must rest upon solid agronomic fact, or its proofs of profitability can be explained away, (as the fertilizer industry is trying to do) as an economic accident brought on by an unusual conjunction of supply shortages and adverse weather. "That agronomic basis," says successful eco-farmer Bill Kurfess, Perry Co., Ohio, "for all of eco-agriculture, is balanced fertility."

Balanced fertility has a very special meaning for

eco-agriculture. It derives from the theories enunciated by agronomists like William Albrecht and biochemists like F. Lyle Wynd thirty years ago. The thrust of their arguments was that if the farmer maintained his soil chemically, biologically, and physically the way nature would do if left alone, then the soil would take proper care of the plants growing on it. "Feed the soil, not the plant" became their watchword, and today's slogan of eco-farming. Albrecht and Wynd severely criticized the fertilizer industry in their day for persuading farmers to fertilize for greater plant yields only, without regard for the biological and chemical balance in the soil, insisting that such practices would lead at worst to toxic soils because of excesses or deficiencies of one or another element, or at best, a costly waste of fertilizer.

Their criticism of "get-rich-quick" schemes of fertility were largely ignored because, let's face it, most farmers, like most any other group of human beings, would like to get rich quick. Ignored, Albrecht's direst predictions all came to pass: soil erosion, the worst pollutant in the nation; fertilizer wantonly wasted as even the fertilizer industry admits; and increasing amounts of chemicals applied to keep yields high. As a result toxic conditions exist in many soils today and in fact may be much more insidious than is now realized. Sadly enough, hardly any farmers got rich quick either.

When a farmer approaches agriculture from the viewpoint of "feed the soil, not the plant" he gauges the importance of soil nutrients by a different yardstick than does the chemical farmer. Nitrogen, rather than occupying first place, drops at least to fifth or sixth place, its position at the top of the list taken by calcium followed by magnesium and potassium in that order. Phosphorus comes next and sodium right behind it. "If you could sum up in a sentence what the key to balanced fertility is," says Dr. Richard

Holliday, president of Eco-Ag, one of the newer eco-farming programs, "it would be the proper ratio between the amounts of calcium, magnesium, potassium, and sodium in the soil," along with trace mineral elements. (Holliday is a doctor of veterinary medicine who quit practicing to join the eco-farming movement, largely influenced by the alarming amount of animal disease he observed which he believes is directly traceable to unbalanced soil. He is also a student of Albrecht.)

To understand why calcium, magnesium, potassium, and trace elements are so important to balanced fertility you need to remember some high school chemistry. Elements, including soil nutrients, are electrically charged and they are able to move in the soil because of electrical attraction. Some are positively charged: calcium, magnesium, potassium, sodium, and most trace elements; some are negatively charged: nitrogen, phosphorus, sulfur. The former are alkaline-forming; the latter acid-forming. The soil particles are also negatively charged, particularly clay whose crystals are so tiny that in any given amount there will be much more attracting surface than in an equal amount of silt or sand. Organic matter, as humus, is negatively charged, too, and both the humus and the clays are called, in this context, *soil colloids*. To negatively-charged soil colloids the positively-charged nutrients are drawn, like metal filings to a magnet, says all agronomic theory, and there they will stay until attracted somewhere else by a stronger "magnet."

Visualize a clay particle in the soil and a plant root beside it,—with my apologies to agronomists for over-simplication. On the negative clay particle are clustered the positive ions of calcium, magnesium, potassium, sodium, iron, zinc, copper, etc. On the root cling positively charged ions of hydrogen. Problem: how are you going to get the plant nutrients over to the root hairs. Calcium is absolutely the most important medium—"It's the key to nutrient availability," says Bill Bailey, Brookside Labs soil consultant in Illinois. "Calcium affects the availability of everything else." Or in the words of Albrecht himself; "Calcium is a 'synergistic' agent. It is a major control of the mechanism which sets up roots so that the other nutrients move into the plant instead of in the reverse direction. . . . Calcium must also be present very early in the life of every plant as can be demonstrated very easily."

Magnesium, too, must get into the plant very quickly, because it is an essential ingredient in the production of chlorophyll.

So how can the root hairs attract the nutrients off the soil colloids? Answer: Sometimes they can't, but when they can it is because the ions on the soil colloids get into geometric positions relative to each other, in which the electrical bond holding them to the clay is weaker than the electrical attraction from the plant root. When and only when the ions of calcium, magnesium, and potassium are present in the proper ratio—balanced with each other, in other words—will the geometrical positions occur that allow flow of the nutrients to the plant roots rather than leave them sticking to the clay.

But let the master speak. Here's how Wynd said it in a famous speech he gave twenty-five years ago and recently reprinted in *Acres U.S.A.*, currently the loudest voice for eco-farming: "The mechanical problem of the escape of a nutrient ion from the vicinity of a colloidal surface in the soil is less if the ions are clustered in certain ratios to each other. It is generally agreed that about 80 percent of the total replaceable bases (that's just another word for the positively-charged ions) should be calcium, about 18 percent should be magnesium, and about 2 percent potassium, to make the proper ratio."

That may not sound like hot stuff to you, but it becomes very controversial when Albrecht uses it as background evidence and explanation for how an insoluble nutrient becomes available for plant food. The conventional agronomists and the fertilizer industry as a whole do not like to talk about "insoluble yet available," which is a contradiction in terms to them. Fertilizers, to be available, must be soluble, they say. Or more exactly, *to be available fast enough to make a quick big yield,* the fertilizer must be soluble. The difference in the two points is more an economic difference than an agronomic difference.

Those figures Wynd gave thirty years ago are not necessarily the exact ones that soil laboratories use today to determine how much calcium or magnesium or potash you may need to apply to balance your soil ecologically. The exact ratio you will use depends upon your particular soil's "cation exchange capability" or "base exchange capability"— both terms mean the same thing. Cation exchange capability (CEC) is a fundamental plank in eco-farming's platform and also one of the scientific planks that links eco-farming and organics so closely together. Cation exchange capability means the amount of nutrient ions a soil colloid is able to hold and then trade to the plant for hydrogen ions in exchange. By extension, the term simply means the amount of nutrients a particular amount of soil can hold. Therefore, a soil with a lot of clay in it and hence more attracting surfaces than sand or silt, will almost automatically have a higher CEC than sandy or silty soil. CEC is measured in units of "millequivalents;" a soil CEC of four to ten ME (millequivalents) you can safely deduce as a sandy soil; if from twelve to twenty ME, a silt loam; if over twenty a heavy clay. As CEC increases, so can the amount of nutrients you add to that soil, but if you add more than the soil can hold, you overload, with resultant waste and/or toxic soil and toxic plants.

That being true, is the farmer with soils of low CEC simply stuck? How can he improve yields ecologically? Answer: By building organic matter and humus, since humus is a soil colloid and by increasing it, you increase the CEC of your soil.

A third fundamental step in eco-farming says that since calcium and magnesium are key *major* elements in eco-agriculture, then lime which is the cheapest and most efficient source of calcium and magnesium, is the most important fertilizer. (Says Neil Broughton, president of Eco-Systems: "The humate industry has posted a $10,000 reward for any agronomist who can prove that nitrogen, phosphorus, and potash are the major plant nutrients. So far no takers. We are thinking of upping the ante. There's no way you can prove that myth. NPK are major only to the fertilizer industry and it is able to control fertilizer labeling regulations so as to keep out competition from other kinds of fertilizer.") *In eco-farming, lime is a fertilizer, not just an acid neutralizer.* Says Albrecht: "The function of lime is not to lessen the acidity itself, but to replace the hydrogen ions absorbed on the soil colloids (when the colloids exchanged nutrients for hydrogen with the plant root) and to disturb the electrical and geometrical symmetry of the ions (of calcium, magnesium, etc.) on the soil colloids. . . ."

Older soil fertility programs following Albrecht, (like Brookside Farms Laboratories, headquartered at New Knoxville, Ohio) are very particular about how the farmers they advise use lime, and especially what kind of lime they use. "Too often farmers just lime as a matter of course to keep pH in the proper range," says Bill Swirbul of Brookside in Jenera, Ohio. "But if you continue to use a dolomitic limestone high in magnesium, you will get an imbalance of magnesium. If you use only a high-calcium lime, you will get too much calcium. Don't keep using lime from the same quarry." Know what's in your lime, and use it accordingly.

Imbalances of one element can quickly lead to imbalances of others. "Excessive calcium causes magnesium, phosphorus, and minor element deficiency," says Ike Falb, president of Viterge Farm Service, Kidron, Ohio. "Excessive potassium, sodium, and magnesium will cause calcium deficiency. Every excess of an element disturbs microflora activity which helps release the huge reserve of minerals bound to silica in the soil."

"When imbalances occur, you get all kinds of problems," says Broughton of Eco-Systems. "Too much magnesium leads to hard, compact, poorly drained land because the clay crystals are more tightly held together for lack of calcium. Tight soil leads to poor aeration which in turn results in anaerobic bacterial action in the soil rather than aerobic. Foxtail and fall panicum and other weeds flourish in an anaerobic soil and these types of weeds are the ones giving chemical farming so much trouble today. In a properly balanced soil you can eventually dispense with herbicides."

"Toxic conditions from excess or deficiency of certain elements result in toxic plants and, eventually, sick animals," says Dr. Holliday. "An imbalance of calcium and phosphorus produces clinical rickets in calves. Selenium must be fed to hogs now in areas where there used to be enough in the soil. Same with iron for pigs. On one of the farms in our program, where soil has been balanced, the farmer no longer has to give his pigs iron shots. Too much nitrogen and not enough magnesium causes grass tetany poisoning in grazing livestock. The important question is, what don't we know yet about the soil imbalances that bad farming has created which may be slowly causing degenerating health in animals and man?"

Balanced fertility in eco-farming is very closely allied to the biological life and to the physical properties of the soil. "As a matter of fact, the chemical contributions to a crop make up only about 15 percent of the whole," points out Bailey. "The biological and physical aspects including how well the soil can make use of water and air are the major contributors." For eco-farming this means emphasis on special minimum tillage and conservation tillage practices (which I will not discuss here), crop rotations, green manuring, all towards the purpose of building and maintaining organic matter. The more organic matter, the greater the soil's colloidal properties to hold nutrients, but even more important, the greater the organic matter content, the higher the population and the more dynamic the activity of soil microorganisms. This activity supplies the plant's needs for nitrogen, phosphorus, and sulfur, the negative ions which are not attracted to soil colloids electrically.

Eco-farming's main, specific difference with chemical farming is over nitrogen. Eco-farming believes that, in the words of Rudolfs Ozolins, soil consultant for Viterge Farm Services: "Nitrogen availability is needed to be *in low concentration and continuous availability* until the plant reaches the ripening stage. Then availability must decrease. All N must be used up, leached out, or built into organic matter. N is a property of the soil's microflora. In some soils, when nitrogen demand sets in, more nitrogen is produced by nitrogen fixation, which is the best form of nitrogen for growing crop plants. If a farmer applies more nitrogen, the demand can never occur and nitrogen fixation can never start. Dangerous conditions can result if N salts or substances are richly available in soil, while other elements, calcium, magnesium, and potassium are deficient. Then the plants take up too much nitrogen for building proteins . . . the stage is set for nitrate and nitrite poisoning."

More or less the same sentiment is expressed by Albrecht and all his followers. Eco-agricultural soil-testing reports always give organic matter content of the soil and a calculation, in pounds per acre, of the amount of slow-release nitrogen the farmer can expect from it. Conventional state soil tests rarely do. Many eco-farmers with good organic matter content need to add no other nitrogen except heavy applications of manure and rotating legumes. That is why the best eco-farms I have seen are those which feed their crops through livestock and have appreciable stores of nitrogen and potassium in the resulting manures to apply to their land.

Many eco-farmers say that an actual soil analysis showing such and such an amount of available nitrogen is not too meaningful anyway, because in a soil high in microorganisms, N is constantly on the move—being absorbed, leached, nitrified, or ammonified. "Analysis of some nutrients in the soil can triple between January and June," says Holliday. "It's an old trick of fertilizer salesmen. Take a soil sample in winter, promise the farmer you can triple his nutrient levels, then take another sample in June, and sure enough you have."

But more important, heavy chemical applications of nitrogen are wasteful and as much as two-thirds yards are not used by the crops at all, say eco-farmers. Nitrogen should be applied as food for microorganisms—the energy they need to reduce carbon compounds to suitable plant food, and let the microbes dole it out to the plants naturally.

Gene Logsdon

The Oxygen-Ethylene Cycle and the Value of Organic Matter

Editor's note: The following article describes a major discovery in soil biology which sheds new light on the protective and beneficial role of organic matter in agriculture. Dr. R. James Cook details a process that he calls the oxygen-ethylene cycle. The cycle is enhanced in the soil when organic matter or compost is added. Its benefits include protection against soilborne diseases, the interruption of the reproductive cycle of plant pathogens, and the stimulation of a variety of hormonal responses by the plant when exposed to ethylene gas produced in the soil.

The benefits of organic amendments to plant health have been recognized scientifically for at least fifty years. One of the first demonstrations by experimentation was in Canada in 1926 when plant pathologist G. B. Sanford showed that potato scab could be controlled almost completely by organic amendments. W. A. Millard and C. B. Taylor confirmed these results only one year later. They worked with green manures that controlled potato scab in Wales. Another historically important contribution came in 1934 from C. J. King and associates in Arizona who reported a very effective control of the dreaded cotton root rot fungus. *Phymatotrichum omnirorum,* using barnyard or other organic manure. Today, it is widely recognized that most soilborne diseases of crops and ornamental plants can be controlled to one degree or another with the incorporation into the soil of a decomposable organic amendment.

Why do organic amendments control soilborne diseases and provide other salutory effects associated with good soil tilth and plant health? This question has been the subject of countless research projects for decades and all over the world. One group of scientists representing the eleven western states organized in 1955 and worked cooperatively for fifteen years on a regional project to explain how crop residues and other organics affected fungus-induced root diseases. A great deal of good basic and practical information came from that project, now summarized in a Washington State University Bulletin, No. 716 (1970) but the project did not answer the basic question—why organics are generally so effective in the control of soilborne diseases.

During a ten-month study leave to Australia, Dr. Alan Smith and I, working at the Chemical and Biological Research Institute of the New South Wales Department of Agriculture in Sydney, discovered a basic soil microbiological process which we believe has far-reaching implications for soil and plant health. The process involves the production of a biologically very active gas, ethylene, which we believe serves as a basic regulator in soil biology.

Ethylene has long been known in plant biology for its influence in ripening fruit, breaking dormancy in buds, triggering rootlet formation, seed germination, and many other hormonally controlled plant growth processes. It was unknown in the soil atmosphere until 1969, when a team of British scientists discovered the gas in anaerobic water-logged soils. Some two or three years later, Dr. Smith discovered small quantities of the gas in well-drained Australian soils, while investigating why the sclerotia (propagative structures about the size of mustard seeds) of the southern blight fungus, *Sclerotium rolfsii*, were inhibited from germinating by non-sterile soil. Dr. Smith showed that ethylene was the inhibitor at concentrations as low as one part per million and less in the soil atmosphere. We now have evidence that ethylene is produced in virtually all soils of the temperate and tropical regions, and that it may be inhibitory to some organisms and stimulatory to others. The reason ethylene has gone undetected for so long is the small quantities present, commonly less than one part per million.

During my visit to Dr. Smith's lab, we concentrated our experiments on trying to determine which organism or group of organisms is responsible for ethylene production in soil. Although there may be several sources, the evidence obtained points clearly to the anaerobic bacteria, which have long been known to be ubiquitous in soils, often in high numbers, but which have heretofore not been ascribed any particularly useful function. These organisms grow only in the absence of oxygen, that is, in oxygen-deficient microsites created for them by aerobic organisms which consume the oxygen.

We have proposed that the anaerobic sites necessary for ethylene production occur with soil crumbs and aggregates. The idea that anaerobic microsites occur commonly in the surface foot of soil is generally accepted, but their importance is only now becoming apparent with the evidence that these are the sites for ethylene production.

Smith and I have further proposed an ethylene/oxygen cycle in soil. As aerobic bacteria and fungi multiply and use oxygen, anaerobic sites develop which permit anaerobes to grow and produce ethylene. The ethylene, in turn, diffuses out from the anaerobic sites and arrests the growth of the ethylene-sensitive aerobes of which there are apparently many, according to earlier work of Smith.

With the reduced aerobic growth, oxygen consumption would presumably also be less, thereby permitting diffusion of oxygen back into the anaerobic sites which arrests growth of the oxygen-sensitive anaerobes. Ethylene production is then reduced, which permits increased aerobic growth, and the cycle repeats.

We also find certain aerobic organisms stimulated by soil ethylene. This might be advantageous to the anaerobes in that, by increasing the rate of aerobic growth around them with ethylene, they could increase the rate of oxygen consumption and thus expand the volume of anaerobiosis needed for their growth and multiplication. On the other hand, by arresting growth of nearby aerobes, they help insure a longer supply of food for themselves that otherwise might be consumed by the aerobes.

Possibly both systems work depending on conditions, organisms present, and food supplies in the soil. In either case, nature's system of checks and balances insures that some ethylene will be produced but also that the producers will not take over the entire soil ecosystem—anaerobes are prisoners of their own microsites.

Where do organic amendments fit in? Of all soil treatments tested to date, organic amendments are the best promoters of ethylene production. Virtually any organic material added to soil stimulates ethylene production, generally within the first twenty-four to forty-eight hours. The reason is two-fold: 1) organic amendments stimulate greatly accelerated aerobic activity and hence increased volume of ethylene-producing anaerobic microsites; and 2) the organic substrates provide the energy-rich food necessary for ethylene production.

In contrast to organic amendments, nitrate nitrogen is the best inhibitor (next to oxygen) of ethylene production found to date. This is apparently because nitrate, like oxygen, keeps the oxidation/reduction potential at a level too high for ethylene production. When nitrate is present, the sequence of microbiological events within a given niche is visualized as follows: 1) oxygen is consumed by aerobes at a rate faster than its diffusion; 2) in the oxygen-free environment, facultative anaerobes (organisms that grow with or without oxygen) use nitrate instead of oxygen in their respiration, converting it to nitrous oxide and even molecular nitrogen (both gases); 3) strict anaerobes begin growth in the oxygen- and nitrate-free site, and produce ethylene. Under even more reduced conditions, methane may even be produced. Thus, it is unfortunate that any nitrates present must first be wastefully discharged into the atmosphere before ethylene production can begin. The ammonium form of nitrogen has no known inhibitory effect on ethylene production, unless first converted to nitrate by the nitrifying of bacteria which are strict aerobes.

We believe a certain level of ethylene is important to soil and plant health. Dr. Smith's work with the southern blight fungus suggests that a level of ethylene in the soil will control this disease. This may be why the sclerotia of this fungus do not germinate even if covered with as little as one-half inch of soil, or also why peanut growers in the south have learned that they must bury the fungus and all plant debris as completely as possible to control southern blight. The soils from two organically managed avocado orchards produce ten to twenty parts per million ethylene within a few days of wetting in laboratory tests. By comparison, soils from *Phytophthora*-conducive orchards produce only one to two parts per million, which may be inadequate to suppress this pathogen. The fact that *Phytophthora* normally inhabits wet soils may mean that it has a limited tolerance to ethylene, thereby requiring higher than normal levels for its control.

The use of mulches to produce ethylene has implications for plant growth as well. Ethylene production is best in warm, moist soils (some ethylene is produced at 50°F. but best production occurs at 80–95°F.). Perhaps plants are triggered to produce new shoots, roots, or even flowers by increased soil temperatures which in turn promote increased ethylene production by anaerobes. The dropping of leaves onto the soil surface may be part of a larger biological process whereby ethylene production is promoted in the natural mulch with warming conditions the following spring, thereby helping the tree or bush through certain growth processes essential in its life cycle. Excessive ethylene causes roots to cease elongation or even to grow upwards. This may help explain the common observation that in forests where the duff is thick and ethylene high, roots grow horizontally through the mineral-rich duff rather than vertically into the more impoverished subsoil.

Ethylene is also a trigger for certain seeds to germinate and thus its proper management has potential in weed control. In North Carolina, ethylene injections are used commercially in sandy, low-organic soils, to trigger uniform germination of witch weed seeds, which are then killed by herbicide. It is undoubtedly significant that no such injections are necessary in soil with higher organic-matter levels. possibly because natural ethylene is adequate to do the job.

Some practices in modern agriculture are harmful to the ethylene-producing anaerobes in soil and thus may have seriously upset a one-time balance between anaerobes and aerobes, and between ethylene and oxygen. The effects of nitrates have already been discussed. Recent work on the use of ammonium with nitrification inhibitors to prevent nitrate formation offers significant possibilities in this regard. Another harmful practice is tillage which destroys the anaerobic microsites, allows faster escape of ethylene and eventually causes a drastic reduction in the ethylene-producing potential of soil. It is interesting that *Fusarium*, an important root rot and wilt pathogen, does not survive in forest soil unless the soil has been intensively tilled. I have consistently found a higher ethylene-producing potential of undisturbed virgin soils compared with adjacent extensively tilled fields of identical soil type and previous history. The current trend towards minimum and even zero tillage for crop production might help reverse this situation. Another practice is the removal of all crop refuse. The continual denial to soil organisms of an adequate and regular food supply greatly lowers the ethylene-producing ability of a soil. In contrast, the incorporation of organic material of various kinds and amounts quickly restores ethylene production.

Soilborne diseases are rare in undisturbed habitats such as grasslands or forests. They are also rare in sites treated regularly with organic amendments. In contrast, these diseases flourish under conditions provided by modern agriculture. Soilborne diseases have been especially troublesome in the west in the once-arid, low organic-matter soil now used for intensive crop production with irrigation. One common thread in this pattern of disease or no disease is ethylene. We think ethylene is an important factor. It may be that some other even more basic factor is involved and that ethylene is only a "barometer" of this more basic process. No possibility should be overlooked. Regardless of which is cause and which is effect, the discovery of ethylene production by anaerobes has opened many new and exciting doors

Organic Farming

for future research on the soil-microorganism-plant ecosystem both for the naturalist interested in biology of wild habitats and for the agriculturalist interested in farm and garden.

R. James Cook
Plant Pathologist,
Washington State University,
Pullman, Washington.
Reprinted from Compost Science.

Natural Fertilizers: How and What are They?

Organic growers are taking a long look at a comparatively new approach to soil fertility; often called "eco-farming," this approach is strong on organic methods. In addition, it has given rise to the manufacture and distribution of a proliferating list of soil amendments that are rocking the staid fertilizer market on its heels. You can now buy all sorts of unusual bacterial yeasts, serum, activators, and conditioners that fall in the area of natural fertilizers. Growers want to know how organic these products are, or even more basically, as one put it: "What in the world are they, anyhow?"

Usage has sanctified the term "natural fertilizers" for these products, but manufacturers are rarely able to call them fertilizers—at least not on the label. Most states won't allow anyone to sell a material as "fertilizer" unless the analysis of nitrogen, phosphorous, and potash is printed on the label. If the analysis of these nutrients is 0-0-0 or thereabouts, as is often the case with soil conditioners, authorities won't concede the product is a fertilizer, even though it may be beneficial to the soil in other ways.

164

If natural fertilizers are so low in the principal plant nutrients, why are sales increasing? Two main reasons: Organic growers who want to expand their operations are looking for sources of organic materials they can buy in quantity and use with greater ease than bulky composts and manures. Second, many chemical farmers have become as concerned with the environment as their organic-minded brothers and are looking for alternatives to chemicals. Third, no small number of farmers, some of them not organically minded at all, at least not in the past, are gravely concerned about the high prices and shortages of chemicals. For reasons of mere survival, they are investigating new methods that do not depend on fossil fuels entirely. As one such farmer succinctly put it: "The (expletive deleted) experts brainwashed us into total dependence on chemicals and then in '74–'75 they either couldn't deliver what we needed or made us pay too (expletive deleted) high a price for it."

Some farmers swear by the natural fertilizers; some swear at them. I won't attempt to name or describe all the products on the market. The ones I do mention are for sake of example. Neither by inclusion nor omission do I intend to show approval or disapproval. For sake of convenience, I divide the materials into four categories: rock product minerals, humates, seaweed derivatives, and bacterial activators. In some cases, materials overlap these categories.

ROCK PRODUCT MINERALS

You're aware of a number of these already, like rock phosphate and granite dust. Calphos is another source of natural phosphate. It's used in organic fertilizer blends where a quick-acting natural phosphate is desired.

165

Hybro-tite is another of the better known rock products organic growers use. Called Lithonia granite, this gneiss rock is mined from deposits in Georgia. It contains about 70 percent silica, 5 percent potash, 14 percent alumina, 1 percent lime, 4 percent soda, 1 percent iron oxide, and traces of many other minor elements. From the organicists' point of view, that 5 percent potash makes Hybro-tite a comparatively good natural source of this essential nutrient. Some growers object to the amount of aluminum in products like Hybro-tite, but the product has had twenty years of satisfied customers. "Hybro-tite is not a substitute for either limestone or rock phosphate, but works well with them," explained Gaines Brewster, sales manager.

Gypsum has been used as a soil conditioner and fertilizer for years—even centuries. Gypsum is calcium sulphate and is one of the key elements in many natural fertilization programs. Yet a significant number of organic growers won't use it. It's natural, maybe, but not organic, they reason. Gypsum is controversial among nonorganicists too: many agronomists consider it worthless on most soils east of the Mississippi even though some farmers on these soils use it in natural fertility programs and claim excellent results.

HUMATES

This group of soil conditioners is making the biggest waves in natural, or eco-farming today. Humates aren't exactly new. The Navahos and Apaches were using them to grow corn in the desert environment of the Southwest at the same time the Delawares were putting fish in their hills of corn in the East. The white man latched on to humates too, but mostly abandoned them when commercial fertilizers became so convenient to use.

Humates are deposits of mineralized organic matter formed much the same way coal was formed. Humates contain high amounts of humus and generally are well endowed with trace or minor elements. Many brand names are on the market and woe to the reporter who suggests that they might all be quite similar. Not at all, insist the distributors.

Eco-Ag Products Corporation describes their product as a glacial humus, deposits of which contain nature's own organic substances and minerals—21 percent to 46 percent organic matter, twenty-five to thirty-six known trace elements, plus bacteria. "What we sell is a blend of plant nutrients, minerals, and humates," says Richard Holliday, DVM, a partner in the Eco-Ag firm at Des Moines, Iowa. The label on a bag of Eco-Ag Plant Food and Soil Conditioner shows an NPK content of 0-2-2— "it's 0-0-0 now," say Holliday—plus 6 percent calcium, 2 percent magnesium, and 1 percent sulphur. The calcium and sulphur may be in the form of calcium sulphate (gypsum), according to Holliday. Also listed on the label are minute quantities of boron, chlorine, cobalt, copper, iron, manganese, molybdenum, sodium, and zinc. For gardens, directions call for ten pounds spread evenly over 200 sq. ft. Cost runs about five dollars per fifty-pound bag, less in larger amounts. Farm cost runs around eight dollars per acre per year, but usually is more than that the first year in the Eco-Ag program.

From Albuquerque, New Mexico, Farm Guard Products sells its soil conditioner, Clod Buster— blended New Mexico humates containing 15 percent humic acids and 30 percent humus, according to the manufacturer, with a pH of 4.5—making it useful in lowering pH in high alkaline soils of the west. "The product is 100 percent organic by any definition," says Leland Taylor of Farm Guard. "Arrowhead

Mills and other organic outlets handle Clod Buster. However, most of our business is improving chemically fertilized soils economically." Some chemical farmers believe humates like Clod Buster prevent leaching of chemicals by building up organic matter, thereby enabling them to cut chemical application appreciably, sometimes by 50 percent.

Wonder Life of Des Moines, Iowa describes its product as high-quality, decomposed, mineralized, natural marine and vegetation humus with added micronutrients and organic protein. Mandrones Mining Co. in Oregon sells a natural soil amendment which it analyzes at 50 percent humus, 37 percent humic acid, and a pH of 7.2, at the opposite end of the acidity scale from Clod Buster. Anvil Mineral Products markets Micro-min, a marine deposit found in Alabama. "It's a heavy clay rich in trace elements," says Ike Falb, Kidron, Ohio, who uses and sells it. Right now the cost is around $250 a ton, but application goes from 200 pounds per acre the first year usually to 100 pounds or 50 pounds every two years. "You can get your cost down to about $6 per acre," says Falb.

SEAWEED DERIVATIVES

Seaweed fertilizers have held high interest for all kinds of growers ever since Dr. T. L. Senn at Clemson University found that seaweed extracts not only produced more vigorous growth, but also seemed to protect plants from some insects.

Glen Graber, Hartville, Ohio, vegetable grower, who started seeking alternatives to toxic insecticides in 1958, achieved considerable insect-protection success with seaweed extracts. Some apple growers have observed that seaweed sprays inhibit the activity of mites, though they don't know why.

Seaweed and kelp extracts are rich in trace elements—Norwegian seaweed containing traces of at least sixty. You can usually buy them in liquid undiluted form or as granules. Seaborn, Maxicrop, Sea Crop, Seaquist are a few of the brand names you might run into.

BACTERIAL SOIL ACTIVATORS

The best-known of these materials, and one at least that has won approval of most (but not all) soil scientists everywhere, is the bacterial inoculant which is applied to legume seeds at planting time by most farmers and gardeners, whether they believe in organics or not. What the inoculant does is make more nitrogen available to the plants. Soybean growers have generally felt, or at least used to feel, that inoculating the seed meant an increase in yield of four or five bushels per acre, at least.

Agriserum is another type of seed treatment that still seems "new" though it's been around for about ten years. It is applied to the seeds directly, where it forms a coating "made up of beneficial bacteria and materials which are nutritional to the seedling and to soil bacteria." "It's an organic serum, completely nontoxic," says Martin G. Aljets, president of Farmers' Manufacturing Co., which makes Agriserum. "We discourage its use with chemicals which seem to inhibit its action. Many farmers are realizing anyway that by growing crops without poisons they get a better food for human and animal consumption."

Agriserum is applied to seeds at the rate of a pint per bushel or about one and one-half pints per acre if seeding rate per acre is more than a bushel. Cost runs about one dollar an acre, though some applications go as high as six dollars.

Canton Mills, Inc. which has been making and marketing its all-organic fertilizer, Shur-gro, since 1956, has a special mix called Shur-gro Activated containing a new blend of nitrogen-fixing bacteria to be used on legume crops. You can also buy Shur-gro in a special high-potash blend, with extra nitrogen and phosphate. "One of the unique features of our organic Shur-gro is that it is all blended with leather," says Delmer Bunke, president of Canton Mills. "No manure, gypsum, peat moss, or any other materials are used in our blends because one does not get enough concentration of plant nutrients with these materials. There's nothing wrong with them; they just aren't concentrated enough for our blends."

Other kinds of soil activators or conditioners use bacteria and bacterial action to increase plant response to nutrients. Medina Soil Activator (Medina Agricultural Products, Hondo, Texas) has been on the market twelve years, a long enough time to denote an appreciable measure of user acceptance. Fertilaid (John C. Porter Enterprises, Austin, Texas) in addition to soil conditioning assets, seems to have the ability to hasten degradation of some chemical pesticides in the soil, according to tests by independent laboratories. Bio-Life 1848, from Bio-Life Co. Inc., Ionia, Iowa, is another bacterial seed and foliar treatment. And of course, most organicists are familiar with Dr. E. E. Pfeiffer's Compost Starter. It is selling now for about $3.50 for thirty grams. That's enough for one ton of material.

There's little agreement among soils experts on the comparative merits of natural fertilizers. Natural fertilizer makers call university agronomists lackeys of the petro-chemical industry (unless one of the lackeys happens to agree with them). The university scientists retaliate by labeling soil conditioner salesman as hucksters selling bags full of magic and hot air.

There is no doubt some truth in both criticisms, but neither side deserves such a blanket indictment. Honest men stand on both sides of the fence. I know university agronomists who have been just as outspokenly critical of some chemical fertilizers as they are of some natural fertilizers. One man in particular stood up against a segment of the petro-chemical industry even under threat of a suit. He didn't think that a particular chemical fertilizer was worth the price it was offered at and he doesn't think some natural fertilizers are worth the money either. I know other agronomists who insist that some farmers are being sold micronutrients they don't need, and the scientists stick to their argument whether the manufacturer is a big chemical company or a small marketer of soil conditioners.

On the other hand, dedicated men also are working in the natural fertilizer industry and they have a vision of what is good in agriculture too. I've known some of them who would stick by a struggling natural farmer when there was no profit in it for them. They sincerely believed that there was another way to produce food, that if farming went solely into the hands of very large chemical farmers, America might be in trouble.

What impresses me as I visit farmers who use the natural fertilizers is not so much the fertilizer itself, but the programs the fertilizer maker insists the farmers follow if they intend to use his product. The best of the natural farmers are getting good results from old-fashioned, nonchemical farming practices: legumes in rotation, erosion control, adequate manure application, minimum tillage operations.

Falb sells Micro-min and other natural fertilizers, so I suppose that makes his opinion a little biased, but he doesn't have to talk about these products when you visit his farm. He doesn't really have to talk at all. You can just look and see. Here are eighty acres only fifty-five of which are tillable, supporting ninety-two head of cattle. Here is lush hay that yields seven tons to the acre on a farm where no chemical nitrogen has been applied since 1965. Here are 125-bushel corn crops and 100-bushel oat crops without herbicides or pesticides. Here is soil on which ten tons of manure per acre is applied annually—soil so good in tilth that a small tractor pulls a four-bottom plow through heavy sod effortlessly. The soil falls over in the furrows in loamy crumbs. You are not surprised to find organic matter content is almost 5 percent! The national average is 1½ percent.

Gene Logsdon

A Scientist Defines an Organic Fertilizer

To begin with, let us recall some basic facts about plant nutrition. Green plants obtain raw materials for their biosynthetic processes in rather simple forms: carbon dioxide, water, nitrate, phosphate, and ionic forms of potassium, calcium, and other essential elements. Nitrogen, to choose a particularly contentious example, almost always enters the roots as nitrate, becoming assimilated by the plant's biochemistry into organic compounds such as amino acids and nucleotides. There is no doubt, then, that nitrate is a "natural" plant nutrient. Nevertheless, a strict organic farmer does not wittingly fertilize his crops with nitrate—or with ammonium salts, which are quickly converted to nitrate by soil bacteria.

Why should a natural plant nutrient such as nitrate be regarded as unnatural when added to the soil as fertilizer? To appreciate this argument we need to go back into soil ecology beyond the immediate entry of nitrogen into the roots. In a natural system, nitrate in the soil is derived from the gradual breakdown of humus, the dark, complex, polymeric material that gives the soil its "tilth." Nitrogen is integrally bound to the carbon atoms that make up the organic structure of humus, which is itself the end product of a complex chain of events that carries nitrogen into the soil. The main path of entry begins with the deposition of organic nitrogenous compounds on the soil in the form of animal feces and urine and the dead remains of animals and plants. These largely organic materials are subjected to hydrolytic and oxidative degradation by decay microorganisms, yielding organic low-molecular-weight products that support the growth of soil microbial flora. These processes finally yield a mass of microbial cells, which, on their death, together with some other remains, become humus. The other source of soil nitrogen is nitrogen-fixation, which also delivers the element to the soil system in organic form. Thus, in a natural soil system, untouched by human technology, nitrogen enters into the system in organic combination with carbon, largely as the nutrient for microorganisms that eventually produce humus.

Now a farmer who wishes to add nitrogen fertilizer to the soil to support crop nutrition has two main alternatives. Nitrogen can be added in a natural, organic form—as plant residues, manure, sewage, food wastes, or, for that matter, in the form of *any* nitrogenous organic compound that can be metabolized by the soil's microbial flora and thereby yield humus. Alternatively, nitrogen can be added in an equally natural, but *inorganic* form, such as nitrate or ammonia. The first choice is the one made by the organic farmer; the second is the conventional route of modern agriculture technology. The strict devotee of natural foods is likely to reject grain grown with inorganic fertilizer in favor of that grown "organically" with manure or compost, sometimes claiming that the nutritional value and keeping qualities are superior—a claim that at this point can neither be confirmed nor denied.

Is there, then, any point in differentiating between the two ways of supplying fertilizer nitrogen? Indeed there is. Considering the soil as an integrated system, there is a vast difference in the outcomes of the two methods. Because nutrient uptake is a work-requiring process, it must be driven by the root's oxygen-dependent energetic metabolism. Humus is much more than a store of nutrients; it is also the chief source of the soil's porosity, hence of its oxygen content, and therefore of the efficiency with which nutrients, such as nitrate, are taken up by the crop.

Thus, the critical difference between the alternative means of supplying nitrogen fertilizer is that the organic form leads to the production of humus, while the inorganic form does not. The use of synthetic urea as a fertilizer provides an informative test of this distinction. Urea is, of course, an authentic organic compound and is, in fact, an ordinary constituent of a clearly natural source of nitrogen—urine. The scientific agronomist may often cite the organic farmer's objection to pure urea as a fertilizer—it is a fairly common one in modern agriculture—as evidence of the irrational basis of organic farming. But is it?

While urea is, indeed, an organic compound, it will not support the bacterial growth that is essential for the formation of humus. When urea is metabolized, the products are ammonia and carbon dioxide. Thus, urea yields carbon in a form that will not support the oxidative metabolism of soil bacteria. To accomplish that, carbon must be in the reduced state, combined with hydrogen, as it is in nearly all more complex organic compounds. Although urea is an organic compound, by failing to support the growth of soil bacteria, and therefore the formation of humus, it does not qualify as an "organic fertilizer."

The intensive use of inorganic nitrogen fertilizer (or urea) may so overload a humus-depleted soil with nitrate as to cause it to leach into surface waters where nitrate levels may readily exceed public health standards. Leached nitrate also wastes expensive fertilizer synthesized from an increasingly diminished supply of natural gas. Apart from any other possible and yet to be established virtues, the use of organic fertilizer (as defined above) avoids these difficulties and holds the promise of restoring the natural source of soil fertility—humus. While it remains to be seen whether food grown in such naturally fertile soil contributes distinctively to the health of people, the practice can, it seems to me, contribute significantly to the health of the soil and the economy.

Dr. Barry Commoner
Director, Center for the Biology of Natural Systems.
Reprinted with permission from Hospital Practice, *Vol. 10, No. 4.*

Chapter 5

Farming with Waste Products

What's Manure Really Worth?

When I was a child on a more or less typical break-even family farm, my mother instilled in me considerable respect for manure. "It's the only profit left after you pay off the bills," she liked to say, drily.

176

I thought it was another of her little jokes about life, (which it was) never dreaming that many years later her words would take on a prophetic ring.

In early 1975 at the University of Illinois, scientists from all over the world gathered at an "International Symposium on Animal Waste" and what the majority of them were saying was that manure has become a valuable commodity. Farmers were listening too, especially beef feeders, because as one of them admitted, echoing my mother years ago: "With fertilizer prices more than tripled and meat prices depressed, the only profit we've had lately is the manure!"

"A dairy herd of 100 cows produces $8,723.50 worth of nitrogen, phosphorus, and potassium (NPK) in its manure," said Dennis F. Meyer, of the Soil Conservation Service at Bismarck, North Dakota. "That means that the 12 million dairy cows in the United States (1973 census figure) are producing $2,868,000 worth of NPK every day, or about $1,046,000,000 a year if we could figure out a way to use it all efficiently."

Before you let that take your breath away, remember that those 12 million dairy cows are but a portion of the total number of livestock, hogs, horses, and chickens in the United States. Estimates vary, but it seems that around 3 billion tons of manure hit the ground every year in this country, which, if given a very conservative value of $3.00 a ton, represents 9 billion bucks. No one at the symposium wanted to be pinned down to that or any figure but it seems about right when compared to Ireland which produces $100 million worth of manure a year, according to Dr. Hubert Tunney of that country. European scientists have such statistics at their fingertips because manure to them is a national resource in a way wasteful Americans have yet to learn. "We have no livestock waste in Ireland, only

Organic Farming

animal manures," said Tunney, chiding those
Americans at the conference who were only seeking
ways to get rid of the stuff.

But trying to determine manure values on the
basis of weight can't be very accurate because
manure varies so much in nutrient content, depending on what the animal has been eating, how much
bedding and/or water has been mixed with the
manure, and how it has been handled and stored.
That's why there's no pat answer to the question of
how much a particular load of manure is worth, at
least not without an analysis of the contents.

Meyer's method of measuring value by NPK
content is more accurate, at least up to a point. The
statistics he uses to arrive at his dollar figures can
also help you determine the value of your homestead
manure.

An average 1,200-pound dairy cow, says Meyer,
following the dairy handbook published by North
Dakota State University, produces 0.6 pound of N,
0.18 pound of P and 0.54 pound of K daily. Meyer
then used January 1975 chemical fertilizer prices to
compute value—N at 29.12 cents per pound; P at
27.83 cents per pound, and K at 11.38 cents per
pound. Therefore a cow produces 23.9 cents of NPK
a day or about $87.00 a year.

Soil scientist D. O. Turner of Washington State
University determines manure value on its nitrogen
content alone. His figures show a 1,000-pound dairy
cow produces .41 pound of N per day. A beef animal
weighing 1,000 pounds produces .34 pound of N per
day, and hogs, per 1,000 pounds of weight, .45
pound of N.

Beef cattle manure (feedlot manure) is not
generally as rich in nitrogen as dairy cow manure,
but it can be. The feedlot manure that Dr. A. C.
Mathers, USDA soil scientist at Bushland, Texas, has
been using in his research, analyzed 30 pounds of N,

178

6 pounds of P, and 20 pounds of K per ton—as rich in N as poultry manure. Even taking into account losses from leaching and volatilization of nutrients, Mathers says that the huge manure piles, once an unsightly liability to feedlots, are now an asset. "With anhydrous ammonia selling at $250 a ton, ten tons of manure (a normal per-acre application) are worth about $18 delivered to the farm." That jibes with what the Texas Extension Service says—a savings of $17 per acre if feedlot manure is used for fertilizer and hauling is not more than ten miles.

But that dollar value of manure may be overly-conservative when measured against the manure's actual field performance. Mather's trials show that manure at ten tons per acre resulted in more grain sorghum per acre *over a five-year period* than did chemical fertilizer applied for maximum yields. The reason no doubt stems from the added organic matter in the manure which results in the cumulative buildup of fertility that organicists have often observed. But you can get too much of a good thing. Mather's trial shows also that regular applications of manure at thirty tons or more per acre causes a decrease in yield and a buildup of nitrates and other salts.

Ray Larson, Kane Co., Illinois feeds out about 3,000 head of cattle and hauls his feedlot manure as a liquid, applying the slurry to his cropland at a 4,000 gallon per acre rate. "Nutrients at that rate are about equal to a chemical fertilizer application of 103-93-146 actual NPK," he says. That makes the manure he hauls worth about $30,000, or $10 per head of cattle.

Many soil scientists won't compare the nutrient content of manure equally with chemical fertilizer when computing value because, they point out, only half of the nutrients in cattle and hog manure are available the first year whereas almost all the nutrients in chemicals can be used by the current crop. Even if this evaluation is justified, a typical northern dairy farm application of twenty tons of manure per acre supplies forty-four dollars worth of nutrients at today's prices, according to Cornell University—*not counting later release of nutrients nor the tremendous value of maintaining good organic matter content.*

Unlike cattle manure, the nutrients in chicken manure are 90 percent available the first year—and chicken droppings are high in N—20 percent to 30 percent. That makes the manure in comparison to chemical fertilizer worth a whopping seven to ten dollars per ton, according to Cornell, and a ten-ton application equal to a twenty-ton application of cattle manure.

A thousand pounds of hogs (five market-size hogs) produce .45 pounds of N per day. That works out to something like $7.00 worth of fertilizer per hog over a five-month lifetime. But much of those nutrients can be lost, depending on how the manure is handled. "Stored in a pit, one pig's waste equals $6.59 in fertilizer value," says A. L. Sutton of Purdue. "If the manure is held in oxidation ditches and agitated, enough nutrients are lost to reduce the value to $4.36 per pig. And if the manure is flushed into an open lagoon, nutrient loss is so high the value of the manure falls to only $1.52 per pig."

"You can't count nutrient value by the amount of excreta," cautions Dale Vanderholm, agricultural engineer at the University of Illinois, "but by how much of the nutrients you get into the soil." Vanderholm has studied all manure handling systems now being used, and he says the one that saves the most nutrients is the time-honored practice of bedded confinement and solid spreading on the field followed by incorporation with plow or disc as soon as possible. Next best is deep-pit storage as liquid with the liquid injected into the soil. The other modern methods—anaerobic lagoons, oxidation ditches, etc., can lose half or more of the nutrients in the manure.

Manure is valuable in ways other than directly as fertilizer. For instance in Georgia, poultry litter is being used to reclaim waste land for agricultural use. Rough hill country is turned into profitable grazing land that can carry a cow per acre by seeding to fescue and fertilizing heavily with chicken manure. "Used to be that a fellow could get a truck load of chicken manure for his garden any time he wanted it," says J. A. Studemann of the Southern Piedmont Research Center, Watkinsville, Georgia. "But now it's hard to find. It's a pretty valuable item."

"In addition to its value as fertilizer, feedlot manure is also a good way to save energy," points out John Sweeten, agricultural extension engineer in Texas. "It takes 5.6 million BTUs of energy per acre to manufacture, distribute, and apply commercial fertilizer at the rate of 180-60-0, while the energy needed to collect, haul, and apply feedlot manure at ten tons per acre is only 1.2 million BTUs."

Another promising use for manure is to recycle it as bedding, mulch, or soil conditioner. For example Babson Bros. (2100 South York Road, Oak Brook, Illinois 60520) has on trial a mechanical system which washes the solids out of the manure so they can be used again for bedding. Or mulch. "We calculate that 100 cows with this system produce $6,687 worth of manure-fertilizer and $3,600 to $4,000 worth of reusable bedding," says A. C. Dale, Purdue ag engineer.

Refeeding manure may become the most valuable way to recycle it. The idea is repulsive to most of us, but it seems to make sense at least chemically. Manure is first of all rich in trace elements (a fact not calculated into its value as fertilizer). In addition, aerobically treated hog manure is richer in amino acids than corn and compares favorably with soybean meal. Trials are being conducted right now, feeding pigs, beef animals, and chickens, hog manure mixed in water (the hog industry euphemistically calls the stuff "oxidation ditch mixed liquor" or ODML).

Processed chicken litter has been fed to beef cattle for about ten years now. "It can replace about one-fourth of beef cattle ration at least," says J. P. Fontenot of Virginia Polytechnic Institute. The value of manure used that way is obvious—over $100 a ton.

The ability of manure to produce methane gas can add considerably to its value especially in warm climates where not much auxiliary heat is needed to warm the manure to 95° for efficient production. What's more, methane production enhances rather than reduces the plant food nutrients in the manure; the gas is truly an added value. The manure from a 1,000-pound beef animal produces thirty cubic feet of gas per day, equivalent to .15 gallon of gasoline or eighteen cubic feet of natural gas. A 1,000-pound dairy cow can produce forty-five cubic feet of gas per day, equivalent to .23 gallon of gasoline or twenty-seven cubic feet of natural gas. A thousand pounds of broiler chickens (about 150 chickens) can produce ninety-two cubic feet of gas per day, equivalent to nearly a half-gallon of gasoline or fifty-five cubic feet of natural gas.

Manure can also be processed into carbon black, building blocks, oil, and other materials. But in every case, either the market is limited or the cost is higher than the final product is worth. For the time being, the American farmer can do no better than utilize his manure for fertilizer and organic matter, as all research reported at the symposium concluded.

Gene Logsdon

There are many different ways to apply sewage sludge. Here it is dumped from a truck and leveled by a Caterpillar, and then plowed into the soil at the end of the day.

Using Waste Safely

Editor's note: This article is adapted from an article in Proceedings of Thirtieth Annual Meeting of Soil Conservation Society of America, San Antonio, Texas, August 10-13, 1975, pp. 160-166.

Agriculture both produces and accepts wastes. Although this relationship has existed since agriculture began, it has become greatly intensified in recent years. The primary reasons are: 1) agriculture enterprises have tended to concentrate and produce large amounts of wastes in localized areas; and 2) increased population, rapidly expanding industrial development, and improved sewage treatment have led to greater amounts of sewage sludge and industrial wastes. These increased supplies of wastes, coupled with the growing concern that wastes must be handled in such a way that the environment is not damaged, have presented real challenges in managing wastes.

There are three primary considerations in utilizing wastes in crop production systems: 1) determining the nutrient content of the wastes; 2) preventing nutrient loss, primarily nitrogen, before incorporation with the soil; and 3) determining the rate that the nutrients in the waste become available for crop uptake.

NUTRIENT CONTENT

Wastes vary greatly in their nutrient content. A chemical analysis of the specific waste to be applied to land is highly desirable. However, this can be estimated from average values of various wastes. Tables 5-1 and 5-2 show some average values and normal ranges of N, P, and K in animal manures, sewage sludge, and sewage effluent. Nitrogen is generally the most important constituent because it is usually the most beneficial to crop production and is also the most likely plant nutrient to cause problems if applied in excess. Excess N applications may lead to nitrate leaching and pollution of water supplies and to nitrate accumulations in forages.

Agricultural processing wastes are also frequently applied to cropland. These wastes, however, are usually low in nutrient value and are of limited benefit, except when they have a pronounced positive effect as a soil conditioner; for example, the spreading of cotton gin trash on sandy soil to control wind erosion.

LOSS OF NUTRIENTS

Solid wastes applied to the soil surface can be lost with runoff water. Also, N compounds can be vo-

TABLE 5-1

N, P, AND K CONTENT OF ANIMAL MANURES (DRY WEIGHT BASIS).[a]

Source	N %	N Range	P[b] %	P[b] Range	K[c] %	K[c] Range
Dairy cattle	3.3	(1.9–5.5)	0.35	(0.1–0.4)	2.0	(1.0–3.0)
Beef cattle	2.0	(1.5–4.0)	0.65	(0.3–0.7)	1.6	(1.0–3.0)
Hogs	4.0	(2.8–7.5)	1.0	(0.2–1.5)	1.2	(0.2–1.6)
Horses	2.5		0.25		0.8	
Sheep	3.5		0.55		1.7	
Broilers	3.5	(1.8–6.8)	2.1	(0.5–3.2)	1.7	(1–2.9)
Laying hens	3.6		1.3		1.3	
Ducks	2.6		0.8		0.5	
Geese	3.3		0.4		0.6	
Turkeys	5.0		0.6		0.8	

[a] Average of values reported by several authors. Manure varies widely in moisture content, with the season, and with the composition of the ration.
[b] Multiply by 2.29 to obtain P_2O_5.
[c] Multiply by 1.20 to obtain K_2O.

TABLE 5-2

N, P, AND K CONTENT OF SEWAGE SLUDGE (DRY WEIGHT BASIS).

Nutrient	Range (%)	Typical (%)
N (total)	1 –6	3
P[a]	0.8–6	1
K[b]	0.1–0.5	0.2

[a] Multiply by 2.29 to obtain P_2O_5.
[b] Multiply by 1.20 to obtain K_2O.

latilized. Both of these potential losses can be essentially eliminated if the wastes are incorporated into the soil immediately after spreading. Robbins, Kriez, and Howells, and Stewart, Mathers, and Thomas have shown that runoff from cropland treated with incorporated manure contained only small amounts of N and P compounds. However, wastes spread on frozen or snow-covered soil were subject to spring thaws without the benefit of soil contact and may lead to large amounts of nutrients in the runoff, In an experiment by Hensler, 10 percent of the N, 6 percent of the P, and 8 percent of the K were lost from winter-applied manure. Minshall, Witzell, and Nichols found a 20, 13, and 33 percent loss of the N, P, and K applied, respectively. Because land slope also affects runoff, some states have issued guidelines restricting waste application on land above a certain percent slope. Maximum percent slope will vary between states because of soil and climatic differences. Also, soil physical factors such as texture, plowpans, and others affecting infiltration rate should be evaluated prior to applying wastes.

Volatilization of N compounds can drastically reduce the value of animal manures and sewage sludges if they are not incorporated with the soil. Salter and Schollenberger summarized fifteen experiments where stable manure was plowed under immediately, six, twenty-four, and ninety-six hours after application. Plots with manure plowed under immediately were given a relative value of 100 in increasing oat yields, and plots with manure plowed under six, twenty-four, and ninety-six hours after spreading had relative values of 79, 73, and 57 percent, respectively. Heck also showed that about 50 percent of the total N in manure could be lost within seven days under warm and windy conditions, and as much as 25 percent could be lost within twelve hours.

RATE OF NUTRIENT AVAILABILITY

The nutrients in animal manure and sewage sludges are not as available for plant uptake as chemical fertilizers. This is particularly true for N, since often only 30 to 40 percent of the total N becomes available during the first year. P and K are considerably more available and often range up to 90 to 100 percent as effective as equivalent amounts of chemical fertilizers.

Smaller amounts of N from manures and sludges become available in subsequent years. This is a very important aspect that requires good understanding to efficiently utilize these wastes. Table 5-3 shows the estimated decay series for several kinds of manure and sewage sludge. The table also shows the ratio of yearly N inputs necessary to supply a constant mineralization rate. For example, 2.5 percent N of dry corral manure has an estimated decay series of 0.40, 0.25, 0.06, which means that at any given application, 40 percent of the N becomes available for plant uptake the first year, 25 percent of the residual N becomes available the second year, and 6 percent of the residual N is mineralized the third and all subsequent years. If ten tons per acre of this manure (dry weight basis) were applied, of the 500 pounds of total N, 200 pounds would be available the first year, 75 the second, 13.5 the third, 12.7 the fourth, 11.9 the fifth, and 11.2 the sixth. The amounts of N mineralized in subsequent years can be similarly calculated.

The ratios shown in Table 5-3 are useful in estimating the benefits from continuous application of wastes. Thus, for the 2.5 percent N dry corral

TABLE 5-3

RATIO OF YEARLY N INPUT TO ANNUAL N MINERALIZATION RATE OF ORGANIC WASTES AT CONSTANT YEARLY MINERALIZATION RATE FOR SIX DECAY SERIES FOR VARIOUS TIMES AFTER INITIAL APPLICATION (PRATT, BROADBENT, AND MARTIN).[a]

Decay series	Typical material[b]	Time, years							
		N input/mineralization ratio							
		1	2	3	4	5	10	15	20
0.90, 0.10, 0.05	Chicken manure	1.11	1.10	1.09	1.09	1.08	1.06	1.05	1.04
0.75, 0.15, 0.10, 0.05	Fresh bovine waste, 3.5% N	1.33	1.27	1.23	1.22	1.20	1.15	1.11	1.06
0.40, 0.25, 0.06	Dry corral manure, 2.5% N	2.50	1.82	1.74	1.58	1.54	1.29	1.16	1.09
0.35, 0.15, 0.10, 0.05	Dry corral manure, 1.5% N	2.86	2.06	1.83	1.82	1.72	1.40	1.23	1.13
0.20, 0.10, 0.05	Dry corral manure, 1.0% N	5.00	3.00	2.90	2.44	2.17	1.38	1.13	1.04
0.35, 0.10, 0.05	Liquid sludge, 2.5% N	2.86	2.33	2.19	2.03	1.90	1.45	1.22	1.11

[a] This ratio equals pounds of N input required to mineralize one pound of N annually.
[b] The N content is on a dry weight basis.

manure, 2.5 pounds of total N in the manure must be added to furnish 1 pound of available N the first year. However, if manure is again added to the same field the next year, only 1.82 pounds must be added to produce 1 pound available, and only 1.54 and 1.09 would be required the fifth and twentieth years, respectively. Consequently, if manure or sludge is added repeatedly, smaller amounts are required each succeeding year. Thus, to provide 200 pounds of available N, ten tons per acre (dry weight basis) was required the first year, 7.3 tons per acre the second, 7 tons per acre the third, and 5 tons per acre the tenth year. Similar estimates for other manures and sewage sludge can be made by using the values in Table 5-3, which are largely estimates of Pratt, Broadbent, and Martin. Although the decay series will certainly vary among locations because of waste composition, climatic conditions, and waste handling differences, these values should be useful in planning waste utilization programs.

HOW MUCH CAN BE USED

For many wastes, the N content will govern the maximum yearly rates that can be applied to land, especially for situations where nitrate leaching may contaminate water or where forages are being grown. Nitrate can accumulate in forages to the extent that feeding them to animals, particularly cattle, can be hazardous.

When large amounts of wastes are applied, decomposition rates may be slower because of the high concentration per unit of soil. The possibility remains that excess nitrate will eventually result, and these systems must be carefully monitored to prevent potential hazards.

Wastes often contain significant amounts of soluble salts. The two salinity problems most often encountered from applying wastes to cropland are excess total salts and high sodium levels. The salt level in rations fed to fattening cattle varies considerably and has a major effect on the salt concentration in the manure. Excess total salts can reduce plant germination and growth. High sodium levels, and potassium to a lesser degree, cause dispersion of soil particles, poor soil structure, and reduced infiltration rates. Therefore, the nature of the salts, as well as the amount, should be determined.

Beef cattle feedyard runoff usually has a relatively high Na concentration and an accompanying high K concentration. A high amount of exchangeable K in soils can have a negative effect on the physical properties of soils, but it is generally regarded as less damaging than Na. To prevent salinizing disposal areas Powers, Clark, Schneider, and Stewart have recommended diluting one part feedyard runoff with at least four parts water. In areas of high rainfall, natural dilution is considerable, and the salinity hazard is reduced.

Large applications of animal manures can also result in salinity problems. Studies at Bushland, Texas, Pratt, Kansas, and Brawley, California, have shown yield decreases from soluble salt accumulations associated with large applications of beef cattle manure. Sewage sludges applied at rates considerably in excess of that required for N fertilizer can lead to salinity problems as well.

Salinity problems can sometimes be prevented by proper timing of waste applications. When feasible, the wastes should be applied several weeks before seeding to allow salts an opportunity to leach from the seed zone. Although salinity problems can result from waste applications, they usually do not persist once waste additions have been stopped, especially under irrigated conditions and in humid areas where substantial percolation occurs to leach the salts from the surface soil.

HEAVY METAL ACCUMULATIONS

Depending on their type and origin, wastes which are available for use on land may contain undesirable levels of trace elements. In contrast with the relatively rapid leaching of salts and nitrate, heavy metals persist in the surface soil and if applied in excess may become a long-term soil management problem. Thus, the decision to use a waste product as an agricultural resource may be contingent on its trace-element content.

If zinc (Zn), copper (Cu), and nickel (Ni) are applied in excess, they are potential threats to soil fertility because of their ability to injure plants. These metals are common in all sewage sludge and refuse and can vary widely depending on industrial waste treatment practices. The potential of Zn, Cu, and Ni to injure crops is very dependent on management factors: soil pH, crop grown, soil cation exchange capacity (CEC), etc. At soil pHs below 6.5 these metals become increasingly available to plants. Levels of metals which severely reduce yields when soil pH is 5.5 cause little or no yield reduction at pH 6.5 to 6.8. Crops differ widely in their sensitivity to excess metals: some crops (beet family, turnips, kale, etc.) are very sensitive to metals; other vegetable crops are more tolerant; corn, soybeans, and small grains are yet more tolerant; and most grasses are even more tolerant. Not only do the crops differ markedly in their general sensitivity to metals, they differ in their relative sensitivity to the individual metals, Zn, Cu, and Ni. For one crop, the sensitivity may be 1:2:4, respectively, while for another crop it may be 1:1:4, 2:1:4, or 1:1:8. Generally, at pH 5.5 to

6.5, Cu is twice as toxic as Zn, and Ni four times as toxic as Zn.

Metals added to soils are quickly bound to different soil or waste components. Ordinarily organic matter, iron and manganese hydrous oxides, and clays bind the plant available metals. Insoluble metal compounds (carbonates, hydroxides, oxides, silicates, sulfides, etc.) may occur in the waste or form in the soil. Because of these metal-binding processes, soils may receive additions of metals up to a certain limit based on soil pH and on other soil properties such as CEC without reducing yields for commonly grown crops. If metals are added to this level for a crop at a pH, subsequent decrease in pH could lead to a crop failure, but this failure could be corrected by liming. Greater metal additions would require change to crops more metal-tolerant.

Recommendations for maximum metal loadings must take into account the relative phytotoxicity of Zn, Cu, and Ni, soil pH effects, crop differences, and soil properties other than pH. Chaney has recommended that no more Zn (equivalent) (the sum of $Zn + 2 \times Cu + 4 \times Ni$) be added than 5 percent of the CEC at pH = 6.5 to allow continued general farming.

Some animal manures may contain as much Cu and Zn as domestic sewage sludge. Baker described the potential for problems from Cu and Zn additions (as antibiotics) to the diets of hogs and chickens when the manures are disposed of on agricultural land.

Protecting the food chain from excessive trace elements is another need during utilization of wastes on land. If uses of a waste would clearly endanger the food chain, it should be discarded; however, current knowledge about soil and crop relationships of potentially toxic elements is inadequate to indicate what is and what is not safe.

Fortunately, most trace elements are held by the soil, or retained in the roots, so plants do not contain dangerous residues. Trace elements are further segregated during fruit or grain formation. Usually, adding excess trace elements to soil leads to crop failure (phytotoxicity) before their residues in the commonly used parts of the plant reach levels dangerous to animals. The major exceptions to this rule are cadmium (Cd), selenium (Se), and molybdenum (Mo). Each of these elements can reach levels dangerous to animals before plants fail (phytotoxicity). Other elements in the food may affect the extent of the injury through interactions.

Industry commonly uses Cd in plating, pigment manufacture, etc., and thus many sludges are high in Cd. Cd disposal into the sewer can be abated by available technology. Because of its potential hazard to the food chain and inadequate knowledge about its soil, plant, and animal relationships, high Cd wastes should not knowingly be added to soils.

Chaney has proposed that Zn phytotoxicity could be used to control the maximum Cd residue in plants. Because most plants are injured when their foliar Zn reaches 500 parts per million, allowing domestic sludges with Cd contents up to 1 percent of their Zn content to be used on cropland would generally lead to Cd foliar contents of no more than about 5 parts per million. Grain and fruit would be minimally impacted. Because abatement of metal release to the sewer could reduce sludges to the 1 percent Cd of Zn criteria, only these sludges should be applied to land. Others would be discarded or used where management could control their food chain impact or used on land which will not impact the food chain.

Other food chain problems can develop from surface contamination of plants by applied sludge or manure or grazing animals ingesting soil. Cows and

sheep commonly ingest soil at a rate equal to 2 to 14
percent of their diet (dry weight). Soils enriched in
trace elements are thus a potential source of impact
on the food chain even though metal-excluding
crops could be grown on the enriched soil.

Considerable research is needed before the major
problems and questions are sufficiently studied so
that utilization of sewage sludge on agricultural land
can be unequivocally adopted. Often sewage sludge
could be used as a micronutrient resource to correct
Zn, Cu, or Fe deficiency. Several additions at N or P
fertilizer levels would supply enough Zn to prevent
Zn deficiency for many years. Perhaps spreading
low metal "domestic" sludges on many farms at low
total metal accumulations would be more beneficial
than using fewer farms at higher total metal accumulations.

B. A. Stewart
Research Leader
USDA Southwestern Great Plains Research Center
Bushland, Texas

R. L. Chaney
Plant Physiologist
Biological Waste Management Laboratory
Beltsville, Maryland

Sludge Safety Guidelines

Most of the conferees at Cornell's Eighth Annual
Waste Management Conference in Rochester, New
York, seemed ready and eager to promote land application of sewage sludge as the disposal method of
choice, rather than as an alternative to dumping or
landfilling. But they wanted some kind of measurement to determine whether sludge is polluted with

heavy metals or is safe to use on land. They got it—from the USDA, not EPA.

Municipal officials who actually give the orders to dispose of sludge on land, wanted EPA to issue a written guideline containing figures and test methods to determine heavy metal safety limits. EPA's Robert Bastian said his agency would not issue written rules. "Within a year or two, the Food and Drug Administration will set seizure levels for heavy metals in agricultural products and we'll be working with those," Bastian said. Seizure levels aren't much help to officials who want to use sludge on land but also want to make certain it's safe.

Dr. Rufus Chaney, the big, outspoken scientist from USDA in Beltsville, Maryland, proposed the rule: the sludge should have a zinc-cadmium ratio no greater than 100:1. At that ratio, he said, the zinc builds up to phytotoxic levels in the soil before the cadmium becomes a danger. In other words, the plants die from zinc poisoning before they accumulate enough cadmium to threaten human health. The zinc-cadmium ratio test is relatively simple to perform, too.

"There's a change at this meeting," said one participant. "People think we've studied sludge too long and now is the time to go out and use it."

The keynote address was delivered by John D. Freshman, a staff member of Senator Ed Muskie's Senate Public Works Committee's Subcommittee on Environmental Pollution. This subcommittee has the primary responsibility in the Senate for developing environmental legislation.

"Land application should not be an alternative disposal method for sludge," he said. "Since it makes the most sense, it should be the first choice. Every land-treatment system I've personally seen has been a success. The problem with sludge use today is institutional, not technical.

"Farmers spend billions of dollars on fertilizers while municipalities spend billions of dollars to dump sewage. Yet the sludge wasted in 1975 alone contained 1,614,000,000 pounds of N, worth $199 million. It contained 1,400,000,000 pounds of P—16 percent of the total used that year—worth $210 million. And it contained 942,000,000 pounds of K, 11 percent of the total, worth $74 million. That's almost a half-billion dollars worth of agricultural nutrients down the drain.

"Land treatment has been studied to death, 'what if'd' to death, but seldom actually done," Freshman said. "Isn't it time to go ahead? Land treatment bears a burden of proof no other waste-disposal system must bear. That's the problem and that's why land application is still considered an alternative. Dumping should be the alternative.

"The law is being ignored. Public law 92-500 calls for land treatment whenever applicable. The intent of Congress in this law is plain: it encourages land treatment. Yet those responsible are dragging their heels. The intent of Congress is being ignored.

"What's needed now is technical assistance to state and local governments to help them get sludge onto the land. Also, graduate schools of engineering should familiarize students with the resource-recovery nature of sewage operations," Freshman said.

Freshman's strong speech drew several supportive comments from the floor. One representative of the University of Wisconsin at Green Bay said that he had recommended land application to his town, but the city built a huge sewage treatment plant. "Now the citizens are stuck with a monstrous tax bill," he said. Freshman commented that he is "very worried over how property owners will be able to pay for those multi-million-dollar sewage facilities."

J. W. Doran of the USDA research station in Lincoln, Nebraska, said that his studies showed that when sludge is put through secondary treatment and disinfected, then used on the land at a rate of five cm effluent per week, only five viable pathogens per square meter of soil were applied. "I conclude that the transmission of disease through sludge is negligible if it's properly treated," Doran said.

S. J. Sedita of the Metropolitan Sanitary District of Greater Chicago gave the results of a four-year study conducted on 15,000 acres in Fulton County, Illinois, land that received 475 dry tons of sludge a day in recent years. "We did surface and groundwater data every month for four years plus we analyzed the physical, chemical, and biological parameters, including the presence of any pathogens or parasites. The data show that public health problems are not significant," Sedita said. "With appropriate safeguards, adequate but not excessive loading, and monitoring, full use can be made of sludge."

Sedita's team measured pathogens in a stream that flowed through the sludged area, taking measurements where it entered the property and where it left. The pathogen level showed no increase over four years anywhere that the water was tested. They also tested two reservoirs. One showed no change in the number of pathogens over four years, in another, the number of pathogens actually decreased slightly. Three wells were tested—two showed no effect and a third, in close proximity to a sludge-storage area, also showed no increase in pathogens. Then his team compared the number of pathogens in sludged soils with control soil from outside the sludged area. "No difference was noticed," Sedita said. Also, animals raised on sludge-grown forage showed no more parasitization than animals fed forage grown on untreated land.

Sludge boosted continuous corn yields about 20 percent over non-sludged plots over an eight-year application period, said Tom Hinesly, a University of Illinois agronomist working at test sites in Elwood, Illinois. He described a feeding study done on pheasants, where they were fed grain grown on fields receiving four different rates of sludge application. On land given no sludge as a control, grain contained .06 parts per million (ppm) cadmium; on land given a total (over several years) of 92 tons of sludge, the rate was .17 ppm cadmium; on land given 184 tons, .33 ppm Cd; and on land given 369 tons, .59 ppm Cd. Even the highest figure was not dangerous, however. The limestone grit fed to the pheasants contained .77 ppm Cd, for instance. The results showed increasing levels of cadmium in livers and kidneys in the birds as the cadmium in the grain increased, but no higher, even at the highest rate, than shellfish or other wild birds. In addition, the flesh of the birds showed no elevation of cadmium at all. It contained amounts similar to wild birds. In one study on swine, the researchers found significant weight gains when the animals were fed grain grown on the land that had received the maximum amount of sludge.

"The only toxicity we've ever been able to produce from sludge is phosphorus toxicity to soybeans using Chicago sludge. I suspect the heavy metals quickly turn insoluble in soils," Hinesly said.

"Would you recommend continued use of sludge, then?" one questioner asked. "Yes I would," Hinesly said, "even with Chicago sludge which is relatively high in heavy metals. At the rate of application of sludge necessary to put adequate nitrogen into the soil, cadmium uptake seems small. There's no reason not to use the sludge, according to my findings."

Dr. Mary Beth Kirkham of the University of Massachusetts at Amherst gave a paper putting the question of cadmium in perspective. A very small amount of cadmium improved chrysanthemum turgor, transpiration, and stomata function. With increased levels, plants did increasingly poorly. But at low levels, she said, cadmium may be essential for plant and animal growth.

"Soil pH is one of the most important factors controlling the availability of heavy metals to plants," she said. "Soil type also has been shown to have an influence on uptake." She then quoted W. T. Sedgewick, writing in *Scientific American* in 1912, to the effect that "there is not one example of sewage farming in the eastern U.S.—it's easier and cheaper for cities to dump. Yet, if Philadelphia's sludge could be brought to New Jersey's pine barrens, Boston's sludge to Cape Cod's sands, New York's sludge to Long Island, then could these barren places be made to bloom." And now in 1976, Dr. Kirkham says, still only a handful of cities in the east are using sludge on the land.

On the relation of organic matter in the soil to heavy metal uptake by plants, she said that the organic matter in sludge binds heavy metals into unavailable forms. "What happens after this organic matter is decomposed, however? Does it then move into plants?" she asked. "Analyses of abandoned sludge sites show no higher concentrations than new sites. We do need more data on the importance of organic matter in controlling the heavy metals, but the organic matter has a binding effect," she said. She added that injecting the sludge deeply into the soil creates anaerobic conditions, and makes more of the contaminants available to plants.

Since shellfish and fish concentrate the heavy metals in their tissues, she feels land is the safest place to dispose of sludge. She said oysters exhibit a bio-multiplication factor of 14,000 for cadmium.

R. C. Sidle, a Penn State agronomist, described the university's Living Filter system in which five cm wastewater per week, carrying a load of 2.5 parts per billion (ppb) cadmium, was delivered to plots of reed canarygrass, corn, and an old field supporting a primary population of white spruce, wild strawberries, and goldenrod. The wastewater was put down every week from 1963 to 1976.

Sidle said that the concentration of cadmium in the reed canarygrass was one ppb and showed no increase from year to year; the cadmium was higher in the control plot in one test. Corn showed an even more surprising result: The Zn/Cd ratio actually decreased from 1.59 percent in 1966 to 0.48 percent in 1975. And there was no increase over the years in Cd in corn tissue. Similarly, the Zn/Cd ratio in the old field was lower than in the control plots receiving no effluent. "Wastewater can reduce the amount of heavy metals in plant tissues because the water creates a greater biomass and the heavy metals are diluted," he concluded.

Dr. Rufus Chaney of USDA, Beltsville, said that his group had been composting sludge and wood chips and using up to 200 tons per acre with no problems.

His group studied the heavy metal content of sludge from thirty cities. The amount of cadmium ranged from a low of 0.6 ppm to 1,230 ppm with a median of 14 ppm.

"Grass and plants grown with composted sludge contain about half the cadmium of plants grown with uncomposted sludge," he said. "Raising the pH with lime seems to hold down Cd uptake, too." These figures indicate that composting reduces the danger of heavy metals in sludge to the point where plants "were not taking up much more cadmium than the control plots," Dr. Chaney said. "I don't know why, but when composted, the cadmium tends to stay in the soil rather than be taken up by plants."

One reason might be that composting raises the pH, and at higher pHs, heavy metals tend to be bound into unusable forms. "Perhaps a reason for this action of composting on heavy metals could be that cadmium and the others oxidize faster in the heat of the pile into unavailable forms," he said.

Chaney said he'd like to see EPA set cadmium and other heavy metals limits for sludge. "The answer," he said, "is to clean up sludge. EPA is too loose on this." Dr. Chaney is working with two cities trying to locate the source of their cadmium pollution.

Since EPA is not going to issue guidelines for sludge use in the near future, Chaney said he is promulgating the following standard:

"A zinc/cadmium ratio greater than 1 percent defines 'polluted sludge'. A Zn/Cd ratio of 1 percent means 100 parts zinc for every part cadmium. At cadmium levels greater than that, at a ratio of 50:1, for instance, the zinc would reach phytotoxic levels before the cadmium became a danger," he said.

"I'd therefore recommend using good sludges agriculturally," Dr. Chaney said.

Jeff Cox

Landowners Attitudes Changing Toward Application of Sewage Sludge?

A sure way to arouse the wrath of rural dwellers has always been to propose spreading city sewage sludge on their farmland, especially if the sludge came from a distant large city. The normal and perhaps justified response was a snarling "No #$%¢* city slicker is going to come out here and dump his $%#$** + on us."

But there's evidence that the farmer, strapped with fertilizer costs that would scare a Rockefeller into receivership, is willing to change his mind on that score—fast—when the disposal operation is carried out prudently.

While a direct land application at Bowling Green, Ohio (the sludge hauled from Toledo), is entangled in a lawsuit over odors, hardly forty miles away in Lima, Ohio, farmers are competing for sludge from that town.

The Lima project did not start out as a lovely bed of roses, either. When first proposed in 1973, the project caused what newspapers euphemistically like to call a furor. Landowners weren't exactly turned on by sewage sludge, but what really raised their ire was the idea of the city taking hundreds of acres of land for the disposal project by eminent domain. Instead of butting heads, however, local government and private citizens got together and in the words of Galen Gault, Utilities Director, "made a responsible effort to find out what the people really wanted and needed."

Result: Some twenty farmers with a total of 4,900 acres have volunteered to have sewage sludge applied by the city. "We actually have more applicants than we can supply sludge to," says Roland Nevergall, waste disposal engineer, who works closely with the farmers in the project. "The farmers have really become receptive to the stuff. The turning point came when they saw the results of some test strips across a corn field before planting. Those strips were up and growing before the other corn. A month or so later, the corn in the strips was a good eight inches taller than the rest of the field."

But a more persuasive reason for the turnaround in farmer attitude was the high price of fertilizer, at least part of which the free sludge can substitute for. "The savings are quite significant," says Nevergall. "Nitrogen content may be as high as 5 percent in our sludge on a dry solids basis and I figure that one application of our sludge (one-half ton of dry solids per acre) is equal to about forty dollars worth of commercial fertilizer. And we try to get on up to four applications if possible."

Robert Manson, at the EPA regional office in Bowling Green, also stresses the value of sludge. "There's no other way to sell land disposal of sludge. If you can't show the farmer how the stuff will benefit him, you might as well quit before you start." Manson points out that in a 1973 survey one community achieved a savings in fertilizer of $41.64 per acre with a seven-ton per acre (dry basis) application of sludge — and since then fertilizer costs have nearly tripled!

That survey, done by Manson and co-worker Clifford Merritt also brought to light other interesting observations. For one thing, the two men found from questionnaires answered and returned by sixty-eight northwestern Ohio communities, that direct land application of sludge was much more acceptable to rural residents who were used to handling and disposing animal wastes on the land. And if the sludge were free, farmers definitely wanted it. "This is particularly true in northwestern Ohio where direct land application has been practiced for a long time. There's no incineration in the area," says Manson. "Take Celina for example. This town of 7,500 has been hauling its sludge to a private farm for twenty years. The farm is very productive and intensely cultivated for a 175-bushel corn crop. In 1973, the commercial fertilizer bill per acre was only $33.95 and even that much may not have been necessary. Agricultural experts agree that with applications of from five to ten dry tons per acre, no additional commercial fertilizer except potassium should generally be needed."

Both plants and soil on that Celina farm have been monitored closely by personnel of the Ohio State School of Agriculture, points out Manson, and their conclusion thus far is that the heavy metals buildup in the soil, now only slightly above background levels, is causing no particular problem but will bear watching for long-term effects.

At Lima with one year's experience in the new direct land application project, Nevergall says, "So far, no problems. I'm able to hold our sludge for thirty to forty-five days during which enough digestion takes place so there is no odor problem when spread. We've had only one complaint about odors and when we inspected the site, we found that what the people were smelling was not the sludge but a nearby application of manure!

"I've just returned from a meeting attended by officials from northwest Ohio communities," added Nevergall. "Looks like most of them are going to go to this kind of waste disposal."

Ohio Farmer magazine has reported favorably on the sludge disposal project—something many farm magazines have been loathe to endorse. It reports that farmers in the Lima project are "well pleased" with the sludge and are conducting comparative trials of their own to see if it improves crops.

What makes the best crop for land disposal of sludge? "Alfalfa is the most desirable," says Manson, "if sufficient nitrogen is available in the sludge to eliminate nodule formation on the roots.

You want the alfalfa to take up as much nitrogen from the sludge as possible to minimize the movement of nitrates into either ground or surface water." Alfalfa is also attractive because it has good value as a crop (hay) and permits additional application after each of three or four cuttings per season.

Corn is a close second as the best crop to utilize sludge, mainly because it's a heavy feeder of nitrogen. Also, grain crops exclude heavy metals from the grain itself, concentrating the metals in leaves and stems. Cadmium is an exception and concentrations of this metal and zinc must be watched closely and evaluated together.

Any grass with good nitrogen utilization is suitable for sludge application. There may be little or no crop to sell to offset the cost of application, but on grass, you can spread when it's too wet to apply with standard vehicles on cultivated soil.

"Spreading is our main problem," says Nevergall. "We have to wait for good weather. We need some of the all-weather transports and big flotation tires like some of the contract haulers have. They can spread every day of the year."

Manson and Merritt give these average disposal costs per ton of dry solids for the various methods: Vacuum filters, centrifuges: $34.41; direct land application of liquid, hauling by contract: $31.93; drying beds: $14.34; direct land application of liquid by city-owned trucks: $7.73.

But Lima is not stopping at just direct application of liquid. At the brand new Shawnee waste-disposal treatment plant on the south side of town, a Zimpro wet-air-oxidation process unit has been installed. Sludge is ground and passed through a heat exchanger where it warms up the incoming sludge while being cooled by it. Then the oxidized, concentrated sludge is pumped into a settling tank where the water is drawn off. The sterile sludge is then ready to be pumped into a truck and spread on the crop field without danger from pathogens. "Our Zimpro unit is one of the smaller ones," says engineer Dave Lloyd at the plant, "but it will handle ten gallons of sludge a minute. We put about 7,000 gallons through it a day."

Farmers are using finished sludge from the Zimpro. Ronald Lloyd applied some last fall on his Elida, Ohio, farm, and doesn't know yet what results he'll get this year. He's had the sludge analyzed by Brookside Farm Laboratories—the consulting service he uses for soil management guidance.

Brookside's analysis of one sample is as follows: N—2.38 percent; P—1.12; P as P_2O_5—2.57 percent; K—.3 percent; K as K_2O—.36 percent. (As is true with most sludge, potassium is not present in sufficient quantity to make sludge a complete fertilizer.) Calcium—3 percent; calcium as calcium oxide, 4.20 percent; magnesium—.94 percent; magnesium as magnesium oxide—1.56 percent; sodium, .09 percent; sulfur—.21 percent; iron—.68 percent; manganese—.03 percent; copper—.043 percent; zinc—.34 percent.

Heavy metals are not a problem at the Shawnee plant. It serves only residential areas. "We had an unusual aluminum content when we first started and which we couldn't explain," says Dave Lloyd. "But it has gone away now."

Gene Logsdon

Chapter 6
Using
Rotations

Why Not Use Chemical N?

Twenty years ago J. I. Rodale spoke at a State Teachers' Convention in Des Moines, Iowa, and was asked by an instructor of vocational agriculture in the audience: "What is the difference between nitrogen and nitrogen?" In other words, why was J. I. making such a fuss back in 1955 whether a plant got its nitrogen in a chemical fertilizer or from animal manure, sludge, or legume?

J. I. loved that kind of question and didn't hesitate a minute in his reply: "Basically, the difference between using nitrogen in pure chemical form or in the form of animal manure is the same as the difference between feeding pure nitrogen to people in pill form as compared to giving it to them in scrambled eggs or peas. I see no other difference."

Throughout the early days of *Organic Gardening and Farming®*, J. I. explained the significance of nitrogen to plant development. Articles always related nitrogen to organic matter and humus: "How Much Nitrogen Does Humus Deliver?," "How Are Nitrogen and Protein Related?," "How Soil Creates Its Own Fertilizer," and "Carbon-Nitrogen Ratio."

J. I. had a knack for simplifying scientific jargon so a gardener and an overworked farmer could quickly grasp the essence and use the findings:

"Organic matter is the means by which most nitrogen is carried in soil. As the organic matter decays, the nitrogen is set free in a form that is available to plants. . . . Adequate nitrogen may be obtained from reasonable amounts of organic matter applied to the soil each year. . . . Nitrogen is necessary to the functioning of every cell of the plant, and is needed for rapid growth. It is directly responsible for the vegetative growth of plants above ground.

" . . . yet an excessive supply of it will not only retard the growth period but will reduce the plant's resistance to disease, and produce crops of an inferior quality, often with poor keeping and shipping abilities. It waterlogs the plant, causing an oversucculency. This is where the organic method is superior. . . . It is a known fact that organic matter decays slowly, thereby not releasing the nitrogen too quickly. The organic matter thus is a valuable storehouse of nitrogen, maintaining an automatic supply for the entire growing season. But chemical fertilizers can sometimes give the soil an oversupply—which will lead to all the troubles described above."

Events since 1955 have added more and more support for organic methods. J. I. Rodale continually warned that scientists were overlooking the excessive use of chemicals in farming and food processing as a cause of cancer. Almost five years after J. I. Rodale's death, a Harvard University professor of atmospheric science warned that "Man's increasing reliance on nitrogen fertilizers may both increase the risk of skin cancer and ultimately threaten the food supply." Testifying before the Public Health and Environment Subcommittee of the House Commerce Committee, in 1975, Dr. Michael McElroy said it was possible that large-scale use of nitrogen fertilizer throughout the world would be a long-term threat to the ozone layer because of the nitrogen compounds that are released.

Elsewhere around the country, from New York's Long Island to the western states, excessive amounts of nitrogen from fertilizer runoff in drinking water present serious health dangers. When the nitrogen count goes to more than ten parts per million, it threatens babies with brain damage and pregnant mothers with abortions. In the last two years research has begun to compare nitrate runoff from farms using organic and chemical nitrogen farming practices. Early test results from the Mike Scully organic farm in Illinois reveal that his fields' nitrate contents remain low (average of 7.9 ppm), while neighboring fields average 49.9 ppm.

The evidence keeps piling up against the one-way route that led away from natural nitrogen sources and straight to the chemical factory. Still to be explored in greater detail is J. I.'s early indictment of declining crop quality: "We can see that something is wrong about the way nitrogen is applied to our crops when we study its effects on the protein in those crops, for protein is 16 percent nitrogen. In a recent ten-year period, the protein content of grain crops in the Middle West declined 10 percent. What else could account for this reduction but the fact that there was something wrong about the nitrogen that was fed to those grain plants?"

While research is finally looking at the bad effects of monolithic artificial nitrogen use, economics lowered the boom. Nitrogen fertilizer prices have gone sky-high. Farmers and gardeners are at last using organic nitrogen sources by the tons and by the bushels.

A typical dairy farm in Pennsylvania, with just forty cows, produces manure that's worth about $1,400 in nitrogen alone, according to Penn State agronomist Wayne Hinish. Phosphate and potash content add another $2,000. Says Hinish: "The total fertilizer value is at least $3,000 and this does not give any credit to carryover fertility (naturally almost 50%) or to the organic matter content. Equally important is the fact that the manure replaced the need for about fourteen tons of fertilizer. It is hard to put a value on this if you can't get the fertilizer."

When J. I. argued the practicality of using cities' wastes as a supplementary nitrogen source, most agronomists asked "Who are you, anyway?" Today agronomists (more than ever before, anyway) boast of 176-bushel corn with sewage sludge applied at ten tons to the acre. Components of dry sludge solids are equal to 3.5 to 6.4 percent N. Tests to determine absence of problems from heavy metals in sludge continue, but results to date look most favorable.

It won't be long until you see advertisements

like this one which appeared in the Burlington, Wisconsin, area newspapers late in 1974: "ATT: Farmers. Beat the high cost of commercial fertilizer, free. The City of Burlington Waste Water Treatment Plant will have available well-digested, odorless, aerobic, activated sludge in a solid state for easy handling. We expect from eight to fifteen tons per day. The nutrient value is as follows: N 5 to 5.6 percent; P 3 to 5.7 percent; K ½ to 2½ percent; and also a very good humus value. . . ."

When J. I. spoke in Iowa, no one talked about the high cost of energy or fertilizer scarcities. No one explained that energy equivalent to twenty gallons of gasoline is needed to make 100 pounds of artificial nitrogen. Just as we have arrived at a new era for manure and other organic wastes, so, too, have we arrived at a new era for legumes.

The weaknesses of the Green Revolution methods which depended upon vast imports of artificial fertilizer, heavy equipment, and automaton-like consultants have been detailed at long last. Today, we hear more about a Green Revolution that makes use of *traditional* farming methods, of *native* resource, of blending the best in science with the techniques of the past.

In this new climate, a Green Revolution using updated organic methods should make significant progress. And in this climate, agriculture in the United States should also improve. We can, for example, learn from the work of N. R. Dhar, a soil chemist at the University of Allahabad.

"I am convinced that the application of nitrogenous fertilizers is ruining the American farms by eating up the soil humus. Without soil humus and organic matter, permanent agriculture is impossible as shown recently by the failure of Dr. Borlaug's 'Green Revolution'."

In a special publication listing the "vital factors in

steady increase in crop production and atmospheric nitrogen fixation," Dr. Dhar reports on his research work in increasing nitrogen content in composts. The type of investigative work undertaken by Dr. Dhar, showing how nitrogen in organic matter can be used more efficiently, is the type of research that should be perfected.

Only a few short years ago, the optimists in food production were the agribusinessmen—the men who pushed petrochemicals. Today, these men are the pessimists, gloomily (at least, on the surface) forecasting shortages in pesticides and fertilizer, while the price of anhydrous ammonia hits $300.

Today, the optimists in food production are the organic gardeners and farmers. And how the world goes about solving the nitrogen question will tell us much about how well-founded our optimism is. If I remember correctly, when J. I. returned to Emmaus after his talk to those Iowa ag teachers, he was convinced that sooner or later they'd understand "the difference between nitrogen and nitrogen."

Jerome Goldstein

Clover: Queen of the Organic Farm

Ask a cornbelt farmer what crop makes him money year in and year out and, without thinking, he'll answer, corn. But if he gives the question more thought, and particularly if he pushes a pencil around the high cost of doing business today, he may change his answer. Legumes, *clovers as well as soybeans*, appear to be the farm's best friends, and certainly must be in any genuine organic operation.

Alfalfa, the most productive of all clovers, already competes well with grains—on the basis of cash cropping only. In 1975, alfalfa in the hands of a good manager easily outpointed corn and soybeans in profit per acre. High-quality alfalfa hay sold for over $100 a ton in parts of the country, and averaged at least $75 a ton everywhere.

At the same time, corn lingered around $2.30 a bushel. Using Michigan State's rule of thumb that you can grow eight tons of alfalfa per acre on any land that will produce 150 bushels of corn per acre, the alfalfa grossed *at least* $600 per acre; corn, $445. And production costs for corn are higher than for alfalfa.

Nor is the price of good clover hay likely to fall lower in relation to grain prices. Every indication is to the contrary. Because of the world protein shortage, any good source of this essential nutrient will remain in good demand. Alfalfa contains as much as 18 to 20 percent *digestible* protein.

"It's the protein angle I use when I want farmers to compare the profitability of corn with alfalfa," says Donald Myers, extension agronomist at Ohio State. "A ton of high-quality alfalfa equals half a ton of soybean meal in protein. When the price of meal is $200 per ton, a ton of alfalfa is worth $100 in protein. With meal at $150 a ton, that's $75 for a ton of alfalfa. On an alfalfa yield of seven tons per acre you're talking about $500 of protein per acre at the lower figure and $700 of protein in the higher!"

Michigan State agronomists put the value of alfalfa in equally graphic terms: four cuttings equaling eight tons per acre has the equivalent in digestible protein *and* net energy (carbohydrates) of ninety-four bushels of corn and seventy-four bushels of soybeans. Using current grain prices, that alfalfa is worth around $600 per acre in feed value alone.

"No other feed performs as well nutritionally for all animals," says Myers, "beef, dairy, swine, horses, even poultry." With enough high-quality hay, dairy cows have produced milk profitably without grain. Beef cattle on well-managed legume and grass pastures will fatten to choice grade with only a fraction of the grain otherwise necessary. Recently, the University of Nebraska carried gestating sows successfully through three farrowings with alfalfa substituting for *all* the grain and soybeans in the normal ration.

Organic farmers, and in fact all farmers who believe strongly in the value of clover, organic or not, appreciate the high value the market is putting on legumes but they will tell you that they would continue to grow clovers in their rotations regardless of the market price. "What my red clover crop saves me in fertilizer justifies growing it even if I didn't make a cent from it in hay or seed," says Dave Haferd, Ohio grain farmer. Other users of chemical fertilizers have found that corn after alfalfa needs only forty pounds of extra nitrogen to produce 120 bushels per acre—on soil that regularly takes over 200 lbs. of nitrogen to produce that kind of yield.

Clovers contribute to soil fertility in many ways, but directly through their unique ability to fix nitrogen in the soil and through their use as green manure. In the first case, the plants draw atmospheric nitrogen from the air, and *Rhizobia* bacteria, living in nodules on the roots of the legume, "manufacture" that nitrogen into a form that plants can use. Traditionally, farmers took advantage of this accumulation of nitrogen by planting corn and other heavy nitrogen feeders after clover.

The amount of nitrogen any particular legume can fix in the soil depends on many factors: climate, region, soil conditions, temperature and available moisture. A USDA bulletin lists the following amounts for various commonly grown clovers:

Alfalfa	194 lbs.	nitrogen per acre
Sweet clover	119 lbs.	" " "
Red clover	114 lbs.	" " "
Ladino	179 lbs.	" " "
White clover	103 lbs.	" " "
Alsike clover	119 lbs.	" " "
Annual Lespediza ...	85 lbs.	" " "
Crimson clover	94 lbs.	" " "
Vetch	80 lbs.	" " "
Pastures with legumes	106 lbs.	" " "

"We don't like to assign specific amounts to any legume," says Myers at Ohio State. "We know that a mature stand of alfalfa will fix between 100 and 200 pounds of nitrogen per acre, depending on conditions. If we measure in April, we may find a 115-pound per acre rate, but by June, the amount may be over 150 pounds. Under the same field conditions, red clover and sweet clover score a little less than alfalfa. But in Ohio we get much less nitrogen from Ladino clover than that USDA table indicates. Farmers have better luck with Ladino in states south of us. In fact, at least for the north, we feel that alfalfa, red clover, and sweet clover are the most practical legumes to consider if your goal is soil fertility."

Apparently there are specific strains of *Rhizobia* for every legume, and it is essential for the best nitrogen production, to inoculate seed with the right strain. If the bacteria are present in sufficient numbers in your soil naturally, you would technically not have to add bacteria with the seed when you plant. "But the cost of inoculant is so low, I advise farmers always to inoculate," says Myers. "Why be half sure?"

In addition to fixing nitrogen, clovers can build fertility effectively as green manure crops. On an air-dry, yield-per-acre basis, the amount of plant material (roots and tops) and the amount of nitrogen contained in the plant materials according to Ohio Extension agronomists is as follows:

	Air-dry Yield per Acre	Nitrogen per Acre
Oct. 1 of seeding year		
Red clover	2,990 lbs.	79 lbs.
Alfalfa	2,750 lbs.	76 lbs.
Sweet clover	4,040 lbs.	121 lbs.
May 24 to June 22 of year following seeding		
Red clover	5,410 lbs.	130 lbs.
Alfalfa	5,490 lbs.	136 lbs.
Sweet clover	6,480 lbs.	155 lbs.

On the basis of green manuring alone, especially when the legume is to be plowed down the fall of the seeding year or the following spring, sweet clover has a slight edge over alfalfa and red clover. However, sweet clover does not fix as much nitrogen nor is it nearly as palatable or nutritious. It can be dangerous as a hay if not properly cured and is not recommended for feeding.

Alfalfa (and to a lesser degree the other clovers) also contains potash—about 2 percent to 3 percent. Which as a green manure can mean as much as seventy-five or more pounds of potash per acre returned to the soil, depending on how heavy the crop.

Summing up, a good clover crop can fix in the soil more than 100 pounds of nitrogen and return another 100 pounds as green manure. That agrees with the old rule of thumb (generally credited to C. G. Hopkins of the University of Illinois back in the thirties) that the amount of nitrogen fixed is approximately equal to that amount found in the hay or mowed portion of the crop. If anhydrous ammonia (82 percent nitrogen) is selling at thirteen cents per pound, as it is right now, (and threatening to go higher) then the dollar value of 200 pounds of nitrogen returned to the soil by a clover crop is equal to twenty-six dollars. Compared to ammonium nitrate at the current thirty cents a pound price, the natural nitrogen is worth sixty dollars per acre. And certainly even the most enthusiastic supporter of chemicals will admit that the clover source of nitrogen is safer and more beneficial to the soil than anhydrous ammonia or ammonium nitrate. Add to the nitrogen value, the dollar worth of the potash at about nine cents per pound today (for muriate of potash), and you have another seven dollars per acre. If you compare the clover potash to commercial organic potash, the value is much higher per pound. All told, a clover crop returns to the soil at least thirty dollars worth of nutrients per acre compared to the cheapest chemicals and double or triple that amount compared to more expensive chemicals or organic fertilizers.

Nor do those calculations take into account any increase in organic matter from the clover. "Actually, I think organic farmers tend to overemphasize the value of a clover crop for organic matter," says Myers at Ohio State. "Certainly you won't decrease organic matter with clover, but the increase will be slight per crop. Only a small percentage of the two tons or so dry matter you plow down remains as organic matter."

Many natural farmers agree. "We think of clover as one of the basic tools of ecological farming," says Neil Broughton, president of N & M Services, which sponsors a natural farming program. "But not for building organic matter as such. There's just not enough bulk in a green manure clover crop. You get more organic matter from cornstalks or from plowing under a heavy dead clover crop in the fall."

But there are other more subtle advantages from clover decaying to humus. "For building tilth, there's no crop like clover," says Bill Kurfess, Ohio natural farmer. "But after the second year, I believe the clover should be turned under. I've not found any gain in building humus by holding a stand any longer than that."

"If you don't use chemicals on it, the clover is great for increasing earthworm populations," says Marion Weaver, another Ohio farmer. "And earthworms can help the clover's soil-enriching program and aerate the soil too."

Soil shielded by a clover sod is almost 100 percent protected from erosion. It will soak up and retain moisture for use in dry times. In fact, clover, and especially alfalfa, is an excellent drought-resistant crop.

Moreover, when soil is conditioned properly to grow a good clover crop—pH nearly neutral, good drainage, and vigorous microbial action in progress from the decaying organic matter, it increases activity not only of *Rhizobia* bacteria, but also *Azotobacter*—a family of bacteria that fix nitrogen freely in the soil, without symbiotic relationship with plants.

Last but not least, clover roots will grow deep into the soil in search of nutrients and bring back to the root zone minerals that have leached too far down for other crops to make use of. In this way, trace elements are returned to poor, farmed-out soils naturally.

Which clover for you? By and large alfalfa, as this article indicates, is the best choice where soil pH is neutral—about 7.0, and drainage good. Alfalfa can be broadcast-seeded in winter wheat or similar crops early in spring, and it can be drilled into a well-worked seedbed in August if moisture is sufficient. But the best results are obtained by drilling it either alone or with oats or other spring-planted cereal grasses in the spring. Chemical farmers seed alfalfa alone and use herbicides to control weeds. But the nurse crop of wheat or oats will also control most weeds. With a nurse crop, the alfalfa stand may not be quite as heavy as without, but you do get the grain crop, so for the organic farmer, it's about an even trade-off. Three or four cuttings can be harvested from alfalfa per year after the seeding year, but the organic grower might be wise to allow his last cutting to die on the stalk and become mulch for over-winter protection of the roots and to put back some potash in the soil as it decays.

Potash is the control factor in alfalfa production. The plant needs a lot of it for high yields, and potash is, of course, the biggest fertility problem for organic farmers. Many natural farmers use refined sulfate of potash which they, unlike strict organic growers, consider organic, safe, and beneficial, despite the small amount of chlorine in it. Some natural farmers favor Shur-gro Leg-U-Blend, a 10 percent potash fertilizer from deposits of kalium in Canada. Kalium needs only very small amounts of hydrochloric acid to make the potassium in it 64 percent available. "Our farmers apply 300 pounds of Leg-U-Blend after the first cutting of hay in summer or sometimes after the last cutting in the fall," says Del Bunke, president of Canton Mills, which makes the fertilizer. "We like to see it used in combination with colloidal clay rock phosphate. Our experience is that livestock will eat 25 percent less alfalfa fertilized this way, but get the same amount of nutrition. Our hay stems, instead of being sort of hollow as is often the case, are solid white inside—made that way by the uptake of phosphorus and potash. We believe this combination of nutrients controls the alfalfa aphid too." The Shur-gro 10 percent potash blend costs about $170 a ton.

Manure is still the best organic source of potash if you can get it in quantity. Where general fertility and organic matter are high, manure at fifteen tons per acre will produce enough potash for a good yield of alfalfa, though ultra-high yields of seven or more tons per acre will need additional potash.

Glenn Graber, natural farmer in Ohio, who also sells Earth-Rite soil conditioner and other organic products, advises greensand at about 500 pounds per acre for additional potash. That's about thirty dollars worth at current prices. Remember that greensand is only about 5 percent potash and not all available right away. "We sow sweet clover in our corn after the last cultivation," says Graber. "Then plow it under the following spring for green manure. We see a great advantage in clovers in that they will take out toxic elements from land that has been chemically mistreated."

"Another advantage of all clovers, at least in our soil, is that they increase percolation and therefore drainage," says natural farmer Virgil Schieber who raises 200 acres of clover for horse hay in Crawford Co., Ohio. "The roots pierce through our tough clay to an underlying gravel layer, and the water can follow the roots down and out. Tile drains work better after a field has been in clover, too."

Floyd Hohman, in Seneca Co., Ohio, likes sweet clover for plowdown because, like most natural farmers, he doesn't use a plow to incorporate organic matter. He uses an offset disc. "Sweet clover stems get stiff at maturity and the disc can cut them up for easier incorporation into the soil even when the plants are five feet tall," he says.

If alfalfa is the best all-around legume clover, and sweet clover the best specifically for quick, short-term green manuring, why do so many farmers still stick with red clover? The first reason is that red clover will grow on wetter soils than alfalfa and will stand a bit more acidity. A good stand will fix nearly as much nitrogen and the plant is not attacked by alfalfa weevil. Red clover can be harvested for seed, too, in all parts of the country, whereas alfalfa can only be practically harvested in drier western states. Yields of red clover seed can run two to three bushels per acre, too, which at a conservative price of sixty dollars a bushel, can mean an excellent income while the crop is enriching the soil. Many midwestern farmers take a two- to three-ton per acre hay crop from red clover in the summer of its second year, then harvest the second crop for seed, then plow down the re-growth for green manure. In a good year, red clover in this kind of program is as profitable as alfalfa.

How do you tell what a potential yield of seed from red clover will be? In a thick stand, if random seed counts range from forty to sixty seeds per head, expect a two-bushel yield. But a one-bushel per acre yield is good and well worth harvesting.

Most clover farmers I know prefer, whenever possible, to use older, smaller combines to harvest clover seed. The straw moves through them more slowly, and wind from the blower can be cut down farther than on many of the new, big combines. More seed goes into the bin instead of back out with the straw.

Seed is harvested when the heads are dead brown and the stems yellow brown in color. A considerable amount of chaff and weed seed is usually still mixed with the seed no matter how good the combining job. You should get the seed spread out to dry in a very thin layer—no more than a couple of inches thick—after combining. Otherwise it will heat and ruin its viability for germination. The seed should be run through a seed cleaner as soon as possible. After that, if drying has been complete (down to 12 percent moisture) the clover seed can be bagged and stored in a cool, dry place where rodents can't get to it. It will remain viable for germination for twenty or more years.

Seed cleaners are still available new. Used ones sell at almost every farm sale in the midwest for around sixty dollars. New ones start at around $300. Almost all grain elevators and feed mills offer seed-cleaning services.

"We've never had potash problems with our red clover," says Schieber. "We just lime and apply rock phosphate. I've seen rock phosphate nearly double yields in alfalfa-red clover mixed hay fields."

Of the other clovers that organic farmers might find useful, Ladino for pasture seedings is certainly one. But Ladino in the north may not be as good as the native white clover which flourishes just about everywhere soil conditions are right for it. White clover and Ladino look about the same except that white is smaller and much hardier. In my fields white clover will spring up and establish a stand all by itself, if the pasture is kept clipped. All it needs is lime. The little clover then begins to fix nitrogen which encourages wild bluegrass to establish itself.

Then the bluegrass crowds out some of the white clover until the extra nitrogen is used up. Back comes the white clover. And so the two plants work together to establish permanent pastures wherever grazing or mowing controls weeds and seedling tree growth.

Crimson clover is a winter annual of the south, producing forage for winter and spring grazing. It will reseed itself but won't survive Northern winters.

Alsike clover will grow on wet ground, and for that reason is sometimes included in pasture and hay plantings.

Birdsfoot trefoil is a long-lasting pasture legume for the north. It, too, will stand poorly drained soils and tolerate soil pH levels too low for alfalfa or sweet clover, but it prefers good soils.

Common and Korean lespedeza make hay, pasture, and cover crops in the middle and southern states but are not often recommended.

Crown vetch and kudzu are both legumes and both (the latter only in the south) have specific application to problem soils. Crown vetch is being used to reclaim strip-mined land where little else will grow. Kudzu is planted on steep grades to control erosion, as is crown vetch on highway embankments.

Whatever your problem, whatever your goal in farming, there's a legume clover to fill your needs. And if you want to know how to operate an organic farm with a profit beyond anything commercial grain growers can achieve, remember that there are a lot of horses in the country now and they all eat hay. And suburban horse owners and race tracks are paying $150 a ton and more in some cases to get it. A word to the wise is sufficient.

Gene Logsdon

The Soybean-Nitrogen Mystery

Soybeans make their own nitrogen and do not seem to need direct applications of chemical nitrogen for good yields. Why this is so is not yet known, but Missouri scientists have made a discovery that might be the key to the answer—with important ramifications for modern organic farmers.

With good reason, the soybean is heralded as the wonder crop of the twentieth century. Whole industries have been built on the protein-packed little golden bean, and new food and feed uses are discovered almost daily.

But the real wonder is that it took us so long to realize the potential of this Asian emigrant. The soybean is at least 2,000 years old; most of its value to human and livestock nutrition was learned in the past 50 years. As late as the mid-1940s, American farmers grew more soybeans for hay than for beans.

Still, knowledge of how to use soybeans has outrun techniques for growing them. More than two-and-one-half times as many scientist's years are spent in soybean research today as ten years ago, and researchers cannot yet say for sure why soybeans grown with no direct fertilizer applications may yield fifty bushels per acre, while those grown with heavy commercial fertilization may yield only twenty-five bushels per acre. Although the chemicals don't inhibit yield, on good soil, they don't add to yield either.

Soybeans are legumes, and fix much of their nitrogen from the air, of course. But that does not completely explain why additional nitrogen applied as fertilizer often fails to boost production—in some cases even seems to limit yields.

We may soon know much more about the growing habits of the soybean, however. A team of researchers at the University of Missouri has a production study underway that promises to unlock many of the keys to growing soybeans. The team, headed by David R. Johnson, agronomist, and Maurice Gebhardt, agricultural engineer, takes an unorthodox approach to research. For instance, rather than tinker around with "micro" research plots and "mini" equipment, the scientists leased a 100-acre farm and bought a line of full-scale farming equipment.

"Eventually, we must fit our research results into realistic production systems that can be used by growers," reasons Johnson. "We decided to do this in experiments where we farm our plots like farmers, rather than hoe out all the weeds, thin to precise stands, and so on. All production practices on this research farm could be used by farmers today."

The practices under study are made from combinations of the following treatments: early May and early June plantings; fifteen- and thirty-inch row spacings; full plow-disc-harrow tillage and no-till; preemergence herbicide and preemerge plus postemerge herbicides. All possible combinations of these practices are used in a total of sixteen different production "systems." In addition, the scientists included areas to study the effect of cultivation, for a grand total of twenty soybean-growing methods.

This "systems" approach to soybean research was started in 1975. The weatherman failed to cooperate: only 2.6 inches of rain fell between June 18 and August 25. As a result, the June-planted beans out-yielded early May soybeans by thirteen bushels per acre. Late August rains came in time to help the June planting, but too late to salvage the May-planted crop.

The researchers found that yields did not differ with tillage method or row width this first year. However, *soybeans cultivated for weed control* out-yielded those soybeans treated with a preemergence herbicide only, and did nearly as well as beans sprayed with a postemergent herbicide following the preemergence weed killer.

With a year's results under their belts, Johnson and Gebhardt are more convinced than ever that their approach is a practical one.

"It's the only study of its kind in the country," says Johnson. "This approach gets everyone working together on soybean production problems. More than twenty scientists are involved in some capacity."

The scientists don't spend all their working hours on tractor seats, however. And the "thinking" time has produced some interesting ideas on soybean culture. To understand some of Johnson's own brainchildren, it's necessary to review how a soybean plant takes in and utilizes nutrients—especially nitrogen.

With most plants, nitrates are taken into the plant from the soil and converted in the leaves to amino acids—the building blocks of protein. Soybeans, being a legume, have a second system. They also take in nitrogen from the atmosphere through nodules on their roots. This nitrogen is converted or reduced to amino acids in the nodules.

"Nitrate reductase is an enzyme that converts or reduces nitrates to amino acids," says Johnson. "Normally, this enzyme operates only in the leaves—that's the case with most plants. But a team of scientists discovered nitrate reductase active in soybean nodules—which indicates to me that nodules, as well as leaves, can convert nitrates to amino acids."

That could explain how big soybean yields are possible without soil applications of nitrogen fertilizer. Depending on soil type, the natural mineralization of organic matter generally makes available 50 to 100 pounds of nitrate per acre per year. With scientific tests, the amount of nitrogen fixed from the atmosphere has been measured at 80 to 100 pounds per acre.

"Where does the rest of it come from?" Johnson has wondered for a long time. "A fifty-bushel soybean crop needs about 300 pounds of nitrogen per acre. At best, nitrogen from the atmosphere and from soil organic matter provides about 200 pounds. We learned long ago that additional nitrogen applied as fertilizer on soybeans is not utilized; does not increase yields. How can a plant get 300 pounds of nitrogen out of a possible supply of 200 pounds or less?"

Tests designed to show how much nitrogen soybean nodules fix from the air indicate only how much amino acid (derived from atmospheric nitrogen) are translocated to the above-ground plant, Johnson notes. Those tests do not show whether some nitrate is converted to amino acids *before* it leaves the nodules. Or whether some other form of nitrogen is utilized by the plant.

"If 1976 research bears out what we've found earlier, it means the nodules are a lot more important to a soybean plant than just to fix atmospheric nitrogen, as we've long believed," says Johnson. "If we can understand this function of soybean nodules, we may be able to apply the knowledge to other crops. Perhaps we could develop a system to provide the nitrogen by biological means for all crops. Nitrogen is the most limiting, most expensive fertilizer element in crop production.

"There are still plenty of questions to be answered," continues Johnson. "We're out in the gray area now—it's about like trying to predict a football score at the opening kickoff. But we've got a lot of research interest in this question of nitrogen fixation right now, and I look for a lot of answers before long."

When those advances in soybean-growing knowledge come, chances are better than good that Dave Johnson's name will be connected with many of them. He's one researcher who asks tough questions—of himself and of his colleagues—and he doesn't accept standard, taken-for-granted answers. When he and his colleagues first reported what he had observed with soybean plants, Johnson found few believers. Now, more researchers are asking some of the same questions he was asking earlier. Do soybeans extract nitrogen from the soil in some form other than nitrate? What really goes on inside a soybean plant?

"I don't know," says Johnson. "But I want to find out."

James Ritchie

Soil Husbandry at Rodale's New Organic Farm

Since early 1972, 290 acres of the New Organic Gardening Experimental Farm have been rented to Ben Brubaker, a neighboring farmer, committed to following organic practices. Through crop rotation and contouring, Ben has worked to overcome the chemicalization of the land that took place before

Rodale purchased the farm. In the process, he has increased the corn yield from 40 bushels per acre (first year after conversion from chemical farming) to a high of 140 bushels per acre in 1975, four seasons later.

Ben uses a "five-year rotation," manuring at about twelve tons per acre once every five years, just before corn's turn in the cycle. But cropping can take two other rotation pathways. For example, after the corn harvest, one third of Ben's total area is planted with soybeans the following summer. In addition to giving the soil good texture, making it light and easy to work, the soybeans fix around 100 pounds of nitrogen per acre. As a result, Ben has often been able to replant this area with corn the next year.

In another third of the acreage, Ben follows corn with alfalfa and oats intercropped. Oats is a nurse crop, and when it's harvested in early July the young alfalfa plants in the stubble are exposed. The alfalfa grows uncut until fall, so it's strong as it goes into winter. Growth is vigorous the next year, and Ben uses it as a hay crop for three, perhaps four, years, usually taking three cuttings a season. Fixing about 200 pounds of nitrogen per acre per year, alfalfa is a deep, strong-rooted plant, but is not as effective a soil texturizer as soy beans because of greater packing by equipment in harvesting hay crops. At the end of this interval, the alfalfa/oats area goes into corn again.

The remaining one third of Ben's acreage goes into a true five-year rotation. After manuring, then planting and harvesting corn, he seeds this area with rye in the fall. When this is harvested in July, the land is readied for an autumn planting of barley. In the fall of the fourth year, wheat is seeded. (To ready the ground for the last step in the rotation—hay—

timothy grass is put in with the wheat.) In March, before the wheat is very tall, Ben seeds in a mix of Mammoth red and white Dutch clovers. Never growing high enough to cut, the Dutch clover is used solely to enrich the soil.

In early summer the wheat is cut, with the hay crop establishing itself during the rest of the season. The next summer, red clover—timothy hay is cut from these fields. Then they are manured over the winter in preparation for the planting of corn that begins the next rotation.

As the three possible rotations suggest, Ben's cropping pattern is complicated. Moreover, in the interest of good soil conservation, his crops are scattered widely over the almost 300 acres in comparatively small plots. Because the hilly land would erode badly if large slopes were plowed all at once, he plants all the crops in small strips to fit the contours, making sure that no adjacent sections are worked up simultaneously. Deeper gullies have permanent grass strips left in them.

Because of Ben's commitment to crop rotation and contouring, his fields offer a striking visual contrast to others around the valley. For example, one rolling twenty-nine acre tract he planted has nine interlocking strips. Parallel strips of barley, rye, alfalfa, beans, and alfalfa run across the side of the high ridge.

At any one time visitors will see about one-half of Ben's land apportioned to small grains; one-fourth to corn; one-eighth each to clover, timothy, and soybeans, and perhaps twenty acres to alfalfa. Ben isn't one to boast, but these three years of work have been repayed with corn production in the higher range for Berks County in 1975. As Ben's satisfaction with this arrangement indicates, organic farming practices are doing a more-than-adequate job of checking erosion and increasing soil fertility at Rodale's New Farm.

Growth-Stimulating Seed Inoculants

Are you getting the most out of the fertilizer you use? Probably not, according to researchers.

In fact, one way to increase farm profits and raise world food production would be to boost the efficiency with which crops use fertilizer.

"Recovery of applied fertilizer nutrients by plants is only about 50 percent for nitrogen and 30 percent for phosphorus," generalized Dr. S. H. Wittwer, director of the Michigan Agricultural Experiment Station at a San Francisco symposium on food and population.

But what are the chances of improving on these fertilizer recovery rates? Pretty good.

Dr. Stanley A. Barber, professor of agronomy at Purdue University, has itemized several possible approaches to increasing nutrient uptake efficiency. One of these is to increase the extent of the plant's root system.

Increasing the number of roots, boosting their growth rate, and expanding the root radius (length of roots) are factors that would increase plant food uptake, he said.

Soil Enterprise Corporation, Stoneville, Mississippi, has developed and begun marketing a line of seed inoculants that show promise of increasing fertilizer efficiency on various crops.

The inoculants basically cause crop plants to develop some of the characteristics outlined as potentially useful by Dr. Barber and other scientists.

In one laboratory test, for example, the Soil Enterprise inoculant increased the length of corn seedling roots by 14 percent and the number of roots by 21 percent. Germination was increased by 2 percent from 66.8 percent for untreated seed to 68.8 percent

Organic Farming

for the treated. The length and number of shoots on the seedlings were increased 25 and 8 percent respectively.

Later during the 1975 season, in a field test of treated versus nontreated corn seed, there was a 13 percent yield advantage for the inoculant. Truckers Favorite corn was planted 2 June 1975 in four replications. Nitrogen fertilizer was applied at the equivalent of 200 pounds of N per acre at lay-by time. Treated plots yielded the equivalent of 60.8 bushels per acre; untreated plots, 53.8 bushels per acre.

What is this inoculant? What does it do?

SOECO Seed Inoculant contains *Azotobacter* bacteria in a medium of peat. Since the early part of the century scientists have explored the usefulness of *Azotobacter*, a bacteria which naturally occurs in soil at rather low populations. Russian scientists have studied the application of this bacteria as a seed inoculant and vast acreages are known to have been treated with *Azotobacter* and other bacteria in Russia for years. Cultures of this bacteria were first made in Russia in 1937 at Leningrad.

Russian scientists said rather early that bacterial fertilization would not replace mineral fertilization. Instead, they felt that the correct combination of bacterial inoculant with mineral and organic fertilizers would enhance plant growth more than any single treatment alone. Other researchers have confirmed this.

Some researchers have had inconclusive findings through the years with *Azotobacter*. This experience has caused a certain degree of doubt about bacterial fertilization. Now the work at Soil Enterprise sheds new light on the practice.

First, as has been found in the past, conditions within the soil affect the survival and functions of the bacteria.

And second, there are many strains of each bacteria. Soil Enterprise has isolated a series of strains of *Azotobacter*, each adapted to a particular crop species. The bacteria of one strain seem to thrive particularly well around the roots of wheat, for example, while the bacteria of another strain are best adapted to the root zone of barley.

The company now markets specific strains for use on wheat, oats, rye, barley, sugar beets, potatoes, and grain sorghum. Also, a cotton strain is being introduced this year. And a mixture of four strains is marketed for use on corn, and a three-strain mixture is sold for sunflowers. A special garden inoculant contains four strains of *Azotobacter* and two strains of *Rhizobium*, the common nitrogen-fixing bacteria.

When seed of a particular crop are inoculated and planted, the bacteria thrive in the exudate which forms around the seed and, within a short time, around the roots. The bacterial colony develops quickly.

Basically, there are two major results:

First, the bacteria produce two key growth hormones, indole acetic acid and gibberellins. These hormones stimulate root growth and development.

Second, the bacteria fix small amounts of atmospheric nitrogen.

This second quality, however, is fundamentally different than that of the *Rhizobium* bacteria on legumes such as soybeans, alfalfa, and clovers. In the case of legumes, the root system has many root hairs, each consisting of one epidermal cell. *Rhizobium* bacteria infect such root hairs and move inside them, forming nodules where nitrogen fixation occurs to the immediate benefit of the plant. This is a true symbiotic relationship—host plant and nitrogen-fixing bacteria mutually benefitting.

The *Azotobacter* bacteria, by contrast, remains outside the plant root in a colony. In this colony there is a constant dying of old bacteria and formation of new bacteria. A single organism may live only a day or so. During this time it fixes only the amount of nitrogen required for its life processes. But when it dies, this organism decays and the cell nitrogen is then taken up and utilized by the plant.

"We believe this technology with *Azotobacter* could be one of the most promising steps toward precision crop production," says Paul Havard, president of Soil Enterprise.

"Up to now, U. S. agriculture has been strongly oriented to chemical solutions. We have tended as an industry to believe that if chemicals couldn't do the job in agriculture, it couldn't be done. At Soil Enterprise we think chemicals will remain important in producing food, but we are equally convinced that soil microbiology has a great deal of promise for the future of agriculture."

Havard became interested in soil microbiology some ten years ago while working as a field representative for a feed equipment firm. Through his readings he learned that European research had identified several soil bacteria as having great promise for inoculation of seed to increase crop yields. Among these was *Azotobacter*.

Subsequently, working with a friend in Germany, Havard obtained a number of *Azotobacter* cultures from that country and organized the research and marketing company now known as Soil Enterprise Corporation.

Company research director Woodrow Blair and staff plant virologist Dr. John R. Yager spearhead the company's research and development program today. The facilities are within shouting distance of the prestigious Delta Branch Experiment Station some ten miles east of Greenville, Mississippi.

Numerous tests of *Azotobacter* strains have been

conducted by Soil Enterprise and by other researchers.

—In a test of field potatoes on a farm near St. Elmore, Alabama, Dr. Bertrand Driskell of the McMillan and Harrison Fertilizer Company, Mobile, produced a 16 percent yield increase with the *Azotobacter* inoculant. Potatoes were fertilized at identical rates in the control and treated rows. Some plots were grown without sidedress fertilizer, others were sidedressed. The average for the whole test revealed that Dr. Driskell produced the equivalent of 207 100-pound bags per acre on the noninoculated rows compared to 241 100-pound bags per acre on the inoculated rows.

"In this test I also tried numerous combinations of fertilizers and micronutrients," explained Dr. Driskell. "The *Azotobacter* inoculant is the only treatment that produced an outstanding increase in yield. I seldom see such an outstanding response when I test new products."

Dr. Driskell said he hopes to put in a five-acre test plot of inoculated potatoes in 1976.

—Wheat yields were increased 14 percent through the use of the inoculant in a company-conducted trial on the Scott Arnold farm near Tunica, Mississippi. Untreated wheat yielded the equivalent of 37.4 bushels per acre; the inoculated wheat yielded the equivalent of 42.8 bushels per acre.

—One location out of four reported a significant increase in wheat yield when the University of Arkansas tested the inoculant around the state. At Manilla, Arkansas, untreated wheat yielded the equivalent of 44.1 bushels per acre; wheat treated at the rate of four ounces per acre yielded 48.7 bushels per acre, for a 10 percent increase. An eight-ounce treatment, however, yielded only 47.3 bushels per acre, for a 7 percent yield increase. And a twelve-ounce treatment rate actually reduced yield from 44.1 in the untreated check to 37.7 bushels per acre.

"We know that many conditions within the soil and even within plants can affect the results of using *Azotobacter* inoculant," said Blair, the company's research director. "And there are clearly some cases where the use of excessive rates of *Azotobacter* reduces plant performance."

—Eggplant responded with a 24 percent yield increase in a company test at Stoneville, Mississippi. Untreated eggplant yielded the equivalent of 16,964 pounds per acre, *Azotobacter*-inoculated eggplant produced 21,090 pounds per acre.

—Oats produced a 138.5 percent yield increase when inoculated with *Azotobacter* in a study conducted by Dr. W. H. Reese of Wayland Baptist College, Plainview, Texas. The test was conducted under extremely dry conditions; only 2.1 inches of rainfall occurred on the test area during the growing season. Treated plants presumably developed more extensive root systems and were, therefore, better able to survive the dry conditions. Plants produced by treated seed weighed almost 77 percent more than those from untreated seed. Yield on the check area was a mere 7.8 bushels per acre; the treated plot produced 18.6 bushels per acre.

—Dr. Reese also measured the response of grain sorghum to inoculation. The weight of seedlings was increased 21 percent by the use of inoculant; the weight of seedling root systems was increased 24 percent.

Some unfavorable results have occurred, of course, as is true in almost all research with new agricultural technology. At Southern Illinois University, Carbondale, nonfertilized corn yielded 104.5 bushels per acre. Inoculated (but not fertilized) corn yielded 93.9 bushels per acre, for a 10.6-bushel-per-acre decline.

In the SIU tests, corn fertilized with 125 pounds of nitrogen per acre yielded 122.2 bushels per acre. With the same rate of nitrogen, plus inoculant, yields increased only 2.7 bushels per acre.

With corn valued at $2.60 per bushel and inoculant for corn at $6.00 per acre, the treatment yielded a net return of only $1.02 per acre (2.7 bu. × $2.60 = $7.02 minus $6.00 = $1.02).

Even in company tests some results have been unfavorable. During 1975, for example, grain sorghum in a field test at Stoneville yielded 233 bushels per acre without any treatment. Where the seeds were inoculated with *Azotobacter*, the yield was 221 bushels per acre, for a 12 bushel per acre decline. In other tests, however, grain sorghum inoculation has produced favorable results.

Several factors can inhibit the action of the bacterial inoculant, and even can cause unfavorable response:

• A German researcher wrote, "If a strong hormone-producer is present, the possibility of overdose exists and the plant growth can be suppressed."

• Adequate moisture is required for favorable growth of the *Azotobacter* colony around crop roots. An English scientist reported that the water content of a soil should be greater than 40 percent of its water-holding capacity.

• Too much water, however, can be very damaging to the colony, since anaerobic conditions will be created, inhibiting the growth of the bacteria.

• Available organic matter also favors multiplication in the bacterial colony, and, therefore, crop response.

• Soil pH must be at a proper level—the organism is said to multiply only when the soil pH is greater than 6. Free aluminum ions found at lower pH inhibit bacterial growth.

• An adequate supply of phosphate in the soil

favors a response from inoculation with *Azotobacter*. When phosphate is limiting, the bacteria may actually compete with plants for the limited supplies.

Farmer and dealer testimonials support the fact that SOECO inoculant can result in profits when the right combination of conditions exists. "I seeded Wells Durum (wheat) on fifth-crop ground," said Don Clark, grain producer and dairyman near Minot, North Dakota. "The SOECO-treated Durum matured five to seven days earlier and yielded forty bushels per acre. Untreated had a lot of 'green heads' when harvested, and yielded twenty-eight bushels per acre." With Durum wheat selling for five dollars a bushel, the twelve-bushel yield advantage for the treated seed meant an additional sixty dollars per acre, which, when considered with the three dollar per acre cost of treatment, means a net profit of fifty-seven dollars per acre.

SOECO distributor Bill Rasmussen, Grand Forks, North Dakota, points out the trend toward greater consideration of technology which complements the use of chemicals for future growth in U. S. food output. Among the many references he has at his fingertips is the June 1975 issue of *Fortune* magazine which contained an article reporting the trend in biological techniques to improve animal and crop performance.

Through his firm, Elk Valley Distributors, Rasmussen has supplied farmers with SOECO inoculant to use in inoculating seed of Durum and hard red spring wheat, oats, barley, sunflowers, corn, sugar beets, pinto beans, and potatoes. Several users during 1975 reported that plants produced with inoculated seed had more extensive, stronger root systems.

"I was particularly impressed with the way the inoculated corn withstood our summer dry spell," said Roy Christiansen, a Minnesota grain producer. The treated corn continued growing to maturity and was harvested as grain; the untreated corn was so severely affected by the drought that Christiansen was forced to chop it for silage.

Growers in North Carolina also have seen very good results from using SOECO inoculant. Joseph Lee, Hollister, North Carolina, saw corn yields increase significantly with the use of the inoculant during 1975. Dealer Quinton Qualls, of Super Yield Soils, Hollister, North Carolina, said his customers used about 125 acres worth of SOECO inoculant on corn and home gardens during the 1975 season. He saw a number of the customers harvest "very measurable yield increases" as a result of the *Azotobacter* going to work in the plant root zones.

Qualls expects to see a vast increase in the acreage treated in his area during 1976, including usage of the product on tobacco. Some acreage of small grains, including wheat, barley, and rye, will be treated, also, said one of his associates, Charles Johnson.

In the carefully controlled laboratory facilities of Soil Enterprise Corporation, where great care is taken to maintain pure strains of the *Azotobacter* organisms, a new element of agricultural technology is, indeed, in development.

Early adopters of new technology who have tried the use of bacterial inoculant have been so impressed that they will try it again. Qualls, the North Carolina dealer, says he hasn't received a negative response from his customers who inoculated their field corn and garden seed in 1975.

William Barksdale

Carefully staked-out plots enabled workers at the Rodale Farm to analyze the effects and benefits of different combinations of interplanting.

Interplanting, How and Why

Interplanting, also called intercropping, is nothing more than growing more than one crop simultaneously on a piece of ground. The advantage is that the technique will increase the land's total harvest. Scientists have found that a field planted to the highest yielding density for one type of plant can still produce a significantly higher total harvest if plants of another kind are grown among the first.

To a farmer used to acre upon acre devoted to just one productive cash crop, this idea will seem revolutionary if not impossible to imagine. But it is im-

portant to recognize that any strangeness in the idea
of intercropping has its roots in the mass production
techniques and equipment, especially for harvesting, which have come to dominate our agriculture.
But these things, which are now the norm for farming all over, were developed to reap the benefits of
mechanization (reducing labor) rather than to utilize
the full potential of the land.

Today the vanguard of agricultural research is
rapidly encouraging a harder look at mixing the
kinds of plants put into a field. Intercropping
represents a feasible way to increase production.
The only extra input it requires are additional seed
and some changes in planting and harvesting
systems. The technique works because it utilizes the
resources in the soil more completely than monocropping.

The mushrooming interest in interplanting really
amounts to a revival. Our forefathers practiced intercropping. We can still see its vestiges in the best
sorts of modern pasture systems. These are always
skillful pairings of grasses and legumes, comple-
mentary in habit of growth and season of harvest.
The legumes are doubly valuable because over the
seasons they enrich the soil's N content while boosting the protein in each crop of hay.

Producing grain and vegetable crops in dense
mixes is still the common practice in cultures which
perform most farmwork by hand. Reports on these
indigenous systems of intercropping have been coming in from agronomists all over. The practice is
truly worldwide and ancient.

The kind of compound advantage, like the one
suggested in the grass/legume pastures, seems to be a
fairly regular effect in a wide variety of crop combinations. The extent of this fortunate and remarkable feature is just now being explored. And of

course the way it works varies widely with the crop plants involved. Here are a few examples of the commonly reported bonuses.

Besides increasing the productivity of soil, water, and solar inputs, interplanting frequently provides a certain amount of pest control. Insect pests seem to be discouraged in several ways. The varied environment encourages more predators, for one thing, and companion plants form natural barriers to epidemic spreading of both insects and diseases.

Denser stands of crop plants mean less place for weeds to grow. And weeds tend to be shaded out earlier in the season, which means less need for cultivation.

These aspects of a farmer's work, in turn, require significantly less labor for maintenance of the fields. And planting two or more crops where there might be just one staggers the expenditure of labor over a longer period which means steadier, more efficient work for people. Then too, when one crop does not yield well a particular season, the companion plant evens out the deficit in the harvest.

The Consultative Group on International Agricultural Research (CGIAR–sponsored by the FAO and the Development Programme of the UN and the World Bank) maintains large research centers around the world. Scientists based at these institutions have most often been the source of the reports on well-established systems of intercropping. Many of their researchers have begun systematic studies to explain intercropping's success. Some are working out ways to increase its efficiency. A look at the work they are doing, as announced in the annual reports of these institutions, will quickly illuminate the scope of intercropping's benefits to food production as well as the state of current research.

African small farmers commonly grow melons,

corn, peanuts, beans, and yams in all sorts of combinations. Agronomists at IITA (International Institute of Tropical Agriculture) in Nigeria have studied a large number of these, confirming the general trend of greater yields. Many kinds of beans will contribute to higher total yields when intercropped with corn. Cowpeas and cassava are another productive pair. Intercropping often discourages the spread of plant diseases, and effectively suppresses weeds especially when viny understory crops are used.

At ICRISAT (International Crops Research Institute for the Semi-Arid Tropics) in Hyderabad, India, interplanting has been found to open a new dimension for native farming by stretching the growing season. Late maturing pigeon peas cropped with a variety of other legumes, sunflowers, or millets produce two crops a year. Waiting to plant conventionally would put the pigeon peas in the middle of the drought. The established system of fallowing the land over the dry season produced no measurable benefit to the crops the following year over interplanting. So not only is a larger crop harvested, but the season of useful employment is also lengthened and the food supply enriched.

Based in the Philippines, IRRI (International Rice Research Institute) has sponsored extensive research on intercropping systems in southeast Asia. Their scientists report high increases in productivity. Corn intercropped with rice yields about 80 percent more than either component monocropped. On Java, various local cropping systems range from 30 percent to 60 percent more productive than sole cropping. IRRI researchers have also demonstrated that intercropping can simplify pest management, for example, by shading out weeds. Mixtures of corn with peanuts and soybeans somehow encourage predatory spiders to move into fields, keeping corn borers in check.

At ICAT (International Center for Tropical Agriculture) in Colombia, the reports again indicate the importance and suitability of beans in intercropping systems. An interesting principle their research turned up is that different varieties of a single type of bean can respond differently to intercropping systems. For example, the yields of a climbing bean grown with corn were reduced far more than the yields of the same variety but of a bush habit. Researchers there are collecting seed from many different sources to use in breeding programs. They plan to develop plant types well suited to intercropping over a range of growing conditions.

Agricultural researchers in the U.S. are just beginning to turn their attention to intercropping. Here where monoculture techniques are so advanced and almost all farmers have made a large commitment to them, there is bound to be resistance. But interest is becoming strong among entomologists, who as a group have been seeking better biological alternatives to chemical insect controls for several years. In the 2 July 1976 issue of *Science*, two biologists describe a model for plant community-insect interaction. Currently Steven Risch of the University of Wisconsin is measuring some of these relationships in the insect populations of monocropped and intercropped fields in Costa Rica. His initial results confirm proportionately lower insect damage in intercropped fields.

At North Carolina State University, Dr. Robert McCollum has completed the first year studying intercrop systems (corn/soybeans and corn/sweet potatoes) which will utilize the three months of growing season available after corn maturity in the South. He has found, like others, that corn is little affected by intercropping. The soybeans yielded 50 percent of normal and the sweet potatoes only 10 percent. He expects to do better by trying different row spacings.

Dr. C. E. Beste of the University of Maryland has had success over three years of research, interplanting sweet corn with soybeans. He has published his study in June 1976 *Horticultural Science*. Several Maryland farmers are already using his system. He observes, "Sweet corn, if grown with soybeans, often appeared dark green and more vigorous than sweet corn alone which suggests perhaps nitrogen fixation by the soybeans may benefit the sweet corn."

Intercropping experiments are being conducted at the New Organic Gardening Experimental Farm. We are probing popular garden vegetable varieties to find combinations responsive to intercropping and the most productive spacings. In all of this we are just beginning to uncover the possibilities. But the beginning we have made is a strong and satisfying one, and is reported on in full in the chapter on Rodale Research.

Jack Ruttle

Chapter 7

Insect and Weed Control

Why Not Chemicals?

What happens to soil life when you use chemical
pesticides and what happens when you use manures
and composts? What effect do the resulting changes
in soil life have on plant health?

Two new scientific books answer these questions with studies which show that life in the soil, when chemically assaulted, loses its ability to feed plants and prevent plant disease, and when organically treated, develops the ability to keep plants well fed and disease-free.*

"The soil flora and fauna together return organic matter to the soil and maintain the carbon cycle," say A. R. Thompson and C. A. Edwards in their paper in *Pesticides in Soil and Water*. "If these organisms are seriously destroyed or affected, then so, ultimately, will soil structure and fertility be influenced. Persistent pesticides have been in use for only about thirty years and the hazard to soil fertility is essentially long-term. It may be some years before further problems become evident," they conclude.

Specifically, they say, "One of the most distinct changes that occurs when pesticides are applied to soil is a simplification of the species complex so that only a few species, often with increased numbers of individuals, remain."

It's the same situation with insects—a hard chemical can wipe out beneficial insects which prey upon crop eaters, allowing a worse outbreak to occur. Or, some insect that was not a problem when the insect community was healthy and balanced, suddenly becomes a major pest when its natural enemies are killed off by chemicals.

Despite a lot of conflicting data, "Most herbicides and insecticides *can indeed destroy soil*

---

*The books are *Pesticides in Soil and Water*, a collection of papers on the interactions of soil life and pesticides, edited by Wayne D. Guenzi and published by the Soil Science Society of America, Madison, Wisconsin; and *Biological Control of Plant Pathogens*, by Kenneth F. Baker and R. James Cook, which shows that soil life, like insect life, consists of mixtures of plant pests and plant friends; published by W. H. Freeman and Co. of San Francisco, California.

microorganisms or suppress their activities if applied at excessive rates . . . these are the acute effects of pesticides on the soil microflora, but little is known about the sublethal or *chronic* effects on soil microorganisms exposed repeatedly to recommended rates over longer periods," says J. F. Parr of the Agricultural Research Service in Baton Rouge, Louisiana, in his paper in *Pesticides in Soil and Water.* He adds that soil fungicides and fumigants cause the greatest alteration in the soil ecology.

Nitrification is one of the most important soil processes from the grower's standpoint—the nitrifying bacteria convert atmospheric nitrogen into compounds that plants eventually use. Farmers are faced with high prices for chemical nitrogen and many of them are turning to manures and legume crops that not only add nitrogen of their own to the soil, but create the right conditions for nitrifying bacteria to flourish.

"Nitrification is one of the most pesticide-sensitive soil microbiological transformations," Parr says. He quotes studies that prove that chlorpropham, an herbicide, inhibits nitrification in direct proportion to concentration. Herbicides amitrole, 2, 4-D and diallate curbed nitrification for at least eight weeks, and another herbicide—propanil—completely inhibited nitrification at fifty parts per million.

Parr ends his paper with a warning. "It is not unusual (for a farmer) to apply to his fields a pre-emergence herbicide, several postemergence herbicides, a preplant fumigant, a systemic fungicide, several foliar insecticides and a defoliant during a single cropping season," he says. He warns that pesticides may interact with one another in the soil to form compounds that may have more serious effects on the soil organisms.

Manure, compost, green manure, and other organic matter are literally food for soil microorganisms, greatly increasing the numbers and disease-preventing powers of beneficial organisms. Proper soil nutrition, which stimulates microorganisms, creates "a state of dynamic equilibrium which is quite stable and has, like the spider's web, remarkable resiliency because it is biologically buffered," say Baker and Cook in *Biological Control of Plant Pathogens,* an enormously informative book that lays the scientific foundation for reasons behind a healthy soil. "This subterranean world is relentlessly competitive, an organism only briefly gaining numerical ascendency, soon to be pushed back into a dormant or inferior condition. Recognition of this led Hepting to comment that 'The old terra firma is a very uncongenial habitat for our plant disease organisms unless they are asleep'. "

Because a healthy soil, fed with organic amendments, is strongly biologically buffered, disease organisms are quickly chewed up and destroyed, or kept dormant and in check, by the already-established beneficial microflora. Baker quotes many studies that show a soil occupied by healthy flora prevents disease organisms from establishing themselves in numbers that can cause symptoms in plants.

But agriculture makes little use of these facts. "The emphasis on killing microorganisms by application of chemicals has been notable in plant disease control . . . and this concept of overkill, unquestioned until recently, still largely prevails," Baker and Cook say. Rather, farmers "must learn to create a stable new biological balance beneficial to their long-term purposes. They cannot restore the primeval balance of virgin soils, but they do have the capacity to direct the establishment of new ones by subtle slow nudging of the flora in the desired direction." The "slow, subtle nudging" is accomplished, the authors say, "through the use of legume cover crops and organic amendments to the soil." They then propose an axiom for soil scientists and farmers to keep in mind, paraphrasing Marston Bates. "The greater the complexity of the biological community, the greater is its stability," and stability means ability to resist an invasion of disease organisms.

Bacteria, fungi, and actinomycetes work together to achieve balance in healthy soil in this way, the scientists say: "Bacteria are generally effective as scavengers and are thus important in competition; a few species may also produce antibiotics. Actinomycetes are poor as scavengers and in competition, but are excellent antibiotic producers. Fungi are effective in competition, hyperparasitism, and some produce antibiotics. Bacteria are effective in the rhizosphere (the zone of soil immediately surrounding plant roots) and all three are effective on the organic debris or crop residues during fallow periods."

Some fungi are predacious and attack nematodes (wormlike soil creatures, some species of which attack plants). According to a study of Duddington and Wyborn done in 1972, these fungi "occur most commonly in soil high in organic matter." "Linford and associates (1938) found that when organic matter is added to soil, the populations of beneficial fungi and free-living nematodes, but not plant-eating nematodes, increase rapidly. The fungi and free-living nematodes then prey on the plant-parasitic forms," Baker and Cook report. Stimulated populations of bacteria "often occur as colonies which coalesce to form an almost continuous cover several bacteria deep on plant roots, according to a 1969 study by Bowen and Rovira. Such bacteria may stimulate plant growth."

"Man has tended, in the past, to use broad-spectrum killers as quick solutions to disease problems, and these, by their very nature, prevent a desirable biological balance, because they simplify the biological community and make it more unstable," Baker and Cook write.

"Although man cannot regain the original biological balance, a new one is possible and must be achieved. Ordish (1967) stated, 'We are only civilized because we have upset the balance of nature to our advantage; unless we continue to do so, we shall perish'. It is becoming increasingly clear that if we continue to accelerate the rate of destroying the total balance, we shall also perish."

Are plants grown in organically treated soil healthier? Baker and Cook tell us exactly what scientists have found out: "Those soils tend to be most suppressive of disease that are highest in organic matter."

In the soil, disease is the exception rather than the rule. Pathogenicity is abnormal and perhaps only occurs when natural biological control is diminished by environmental disturbance (such as the application of certain chemicals).

"Man must learn to visualize the pathogen on his crops as a partner at the feast, there before himself, and in the overview, as much a part of the scene as he, and more likely to hold a residence permit. Each organism is as much the center of its own universe as man believes himself to be," Baker and Cook conclude.

"Man's attempt to feed his teeming multitudes frequently disturbs the delicate balance below ground, as well as above. His naive assaults on the subterranean biological network often result in his entrapment; in freeing himself from one strand he becomes entangled in others. Numerous fortuitous demonstrations have shown that agriculture can function within the ecological limits of the soil flora and fauna, and these examples man must study deeply and reflectively. Unable to win by annihilating all competitors below ground, as he can the animals and plants above, man must come at length to understand the effects of what he does, and work with, rather than ignore, the established order of the earth."

Realities of a Pesticide Ban

Agricultural pesticides and their side effects have received so much publicity during the past ten years that it would be easy to assume first, that the chemicals are used on all farmland in the U.S., and second, that agricultural productivity is completely dependent upon the pesticides. Data discussed below show that both assumptions are wrong. As to the first, only a small proportion of farm acreage is treated with pesticides. For example, just 5 *percent* of the nation's total crop acreage receives insecticide applications. As to the second assumption, the overall supply of food crops, except for some fruits and vegetables, would not be seriously diminished if pesticide use were substantially reduced. Total annual loss of crops from pest damage would rise only about seven percentage points, from 33.6 to 40.7 percent, if pesticides were eliminated. In general, the increased loss in most cases could be compensated for by increased planting, crop substitutions, and reemphasis on a number of conventional agricultural procedures. Furthermore, even with intensified pesticide application, some crop losses now are rising because of such insecticide-related factors as acquired resistance, destruction of natural enemies, and emergence of formerly nonpest insect species.

Much of this study is based on U.S. Department of Agriculture (USDA) survey data. Some of these data were collected in various studies during the 1960s; unfortunately, more recent data are not available. Furthermore, some of the information is based on estimates gathered in surveys; thus there are inherent limitations, but the data are the most complete available. Despite these reservations, such a study is worthwhile, since too many claims and counterclaims concerning the risks and benefits of pesticides have been made with little, if any, attention given to available data, however incomplete.

Nearly a billion pounds of pesticides, or about five pounds per person, are used annually in the U.S. Of the one-half billion pounds of pesticides applied to crop and farm lands, 54 percent are insecticides, 36 percent herbicides, and 10 percent fungicides. Nearly another half billion pounds of pesticides are used by government agencies, industries, and homeowners. The following analysis considers only those amounts used for agriculture.

Pesticide use in agriculture (Table 7-1) is not evenly distributed; for example, 50 percent of all insecticides used in farming is applied to the nonfood crops of cotton and tobacco.

Of the food crops, corn, fruit, and vegetables receive the largest amounts of insecticide. Of the herbicidal material applied, 41 percent is used on corn, the remaining 59 percent on the other crops. Most of the fungicide is applied to fruit and vegetables, with only a small amount used on field crops.

Despite increased use of pesticides, only 5 percent of all cropland in 1966 was treated with insecticides, 12 percent with herbicides, and 0.5 percent with fungicides. If the amount of land devoted to pastures is removed from the total of 890.8 million acres of cropland in the U.S., the percentage of cropland (including that planted with nonfood crops of cotton and tobacco) treated with insecticides, herbicides, and fungicides is larger: 12 percent, 27 percent, and 1.3 percent, respectively.

Although cotton receives nearly 50 percent of all insecticides used in agriculture, about half (46 percent) of the total cotton acreage receives no insecticide treatment at all. The largest percentage (79 percent) of the acres treated is in the southeastern and delta states, whereas the smallest percentage (37 percent) treated is in the southern plains. Of the food crops, only citrus, apples, and potatoes have more than 85 percent of their acreages treated with insecticides. Many acres of small grains and pastures receive little or no treatment with insecticides.

Average insecticide treatment for all crops equals 1 percent of the crop acreage in the mountain-states region, 3 percent in the northern plains, 17 percent in the corn belt, and 19 percent in the Southeast. This wide range is in part due to differences in crops grown, intensity of insect attack, and cultural practices followed in the different regions. In addition to these variations, pesticide treatment for any one crop may differ according to geographic region. For example, in the northern plains, where large quantities of potatoes are grown, 42 percent of the potato acreage is treated with insecticides, while in the Southeast, where early potatoes are grown, 100 percent of the potato acreage is treated. This difference probably reflects the higher intensity of pest attack occurring in the warmer regions.

Herbicides for weed control are applied to only 12 percent of the total crop acreage (Table 7-1). Of all pesticide applications, however, those for herbicides have increased the fastest. Those crops which have more than half of their acreage treated with herbicides include peanuts, corn, potatoes, cotton, and rice.

In general, the large growers (with sales of $40,000 or more) applied more pesticides to a larger per-

TABLE 7-1
PESTICIDE USAGE IN AGRICULTURE

Specific Crops	Percentage of total U.S. crop acreage	Insecticides		Herbicides		Fungicides	
All U.S. cropland		Percentage of all cropland treated with insecticides **5**	Percentage this use is of total pesticides used on all cropland **54**	Percentage of all cropland treated with herbicides **12**	Percentage this use is of total pesticides used on all cropland **36**	Percentage of all cropland treated with fungicides **0.5**	Percentage this use is of total pesticides used on all cropland **10**
		Percentage of all acres farmed (for indicated crop) treated with insecticides	Percentage of total U.S. insecticide use in agriculture devoted to indicated crop	Percentage of all acres farmed (for indicated crop) treated with herbicides	Percentage of total U.S. herbicide use in agriculture devoted to indicated crop	Percentage of all acres farmed (for indicated crop) treated with fungicides	Percentage of total U.S. fungicide use in agriculture devoted to indicated crop
NONFOOD	1.26	1	50	0.5	NA*	less than 0.5	NA
Cotton	1.15	54	47	52	6	2	1
Tobacco	0.11	81	3	2	NA	7	NA
FOOD	98.74	4	NA	11.5	NA	less than 0.5	NA
Field Crops	NA	NA	NA	NA	NA	NA	19
Corn	7.43	33	17	57	41	2	NA
Peanuts	0.16	70	NA	63	3	35	4
Rice	0.22	10	NA	52	2	0	NA
Wheat	6.11	2	NA	28	7	0.5	NA
Soybeans	4.19	4	2	37	9	0.5	NA
Pasture	68.40	0.5	3	1	9	0	NA
Hay and Range							
Vegetables	NA	NA	8	NA	5	NA	25
Potatoes	0.16	89	NA	59	NA	24	12
Fruit	NA	NA	13	NA	NA	NA	NA
Apples	0.07	92	6	16	NA	72	28
Citrus	0.08	97	2	29	NA	73	13

*Not available.
Sources: Extent of Farm Pesticide Use on Crops in 1966. Agr. Econ. Rep. No. 147, USDA Econ. Res. Ser., 1968. Quantities of Pesticides Used by Farmers in 1966, Agr. Econ. Rep. No. 179, USDA Econ. Res. Ser., 1970.

Organic Farming

centage of their acres than did the small growers (with sales less than $2,500). The percentage of land treated (excluding pastures) ranged from 6 percent for the small producer to 21 percent for the large producer. Some individual crops, however, provided exceptions to those trends, potatoes in particular.

According to the latest USDA estimates (1951 to 1960), crop losses due to all pests were $9.9 billion, or 33.6 percent of the total crop. Losses due to insects and plant diseases have increased since the previous decade, but losses from weeds have decreased significantly.

The use of DDT and the other synthetic insecticides has increased since their introduction in 1946. Although overall crop losses from insects have gone up despite the use of insecticides, important reductions have been made in losses caused by certain pests in some crops. For example, losses from potato insects declined from 22 percent in the period from 1910 to 1935 to 16 percent in the period from 1942 to 1951, and to 14 percent in the period from 1951 to 1960. This reduction was expected in view of the effectiveness of insecticides in controlling the major potato insect pests.

In contrast, losses in apples, caused primarily by the codling moth and the apple maggot, generally have not declined with increased use of organic insecticides. A 10.4 percent loss was reported for the period from 1910 to 1935, a 12.4 percent loss from 1942 to 1951, and a 13.0 percent loss for the period from 1951 to 1960. This loss pattern probably reflects higher appearance standards for salable fruit as well as the decline in cultural and sanitation controls formerly practiced in orchards to counteract these pests.

According to USDA estimates, corn losses due to

insects have been increasing. A 3.5 percent loss was reported for the period from 1942 to 1951 and a 12.0 percent loss for the period from 1951 to 1960. Factors contributing to increased corn losses due to insects include continuous cultivation of corn on the same land year after year and the planting of insect-susceptible rather than resistant types. This latter factor has been implicated in the greater losses in rice and wheat varieties used in the "green revolution."

Estimated crop losses if no pesticides were employed are presented in Table 7-2. Without any pesticides, crop losses from insects would increase to $4.8 billion (16.3 percent), losses from diseases to $4.2 billion (14.2 percent), and losses from weeds to $3 billion (10.2 percent). Total losses without applying any pesticides are estimated at $12 billion, an increase of $2.1 billion over the losses when pesticides were used during the period from 1951 to 1960. Eliminating pesticides would increase crop losses to 40.7 percent of potential crop production. This is an increase of only 7.1 percentage points over the losses incurred *with* use of pesticides from 1951 to 1960.

These estimated crop losses are statistically exaggerated, with or without insecticide application. The reason is that insect, disease, and weed losses were assessed separately and then added together. For example, insect and disease attacks on one apple were counted as a loss both from insects and from disease. This approach yields an estimated total loss for apples from insects, diseases, and weeds of 150 percent (proportionately: insects 60 percent, disease 80 percent, weeds 10 percent). Obviously, total apple losses cannot be greater than 100 percent; a more accurate estimate is a 90 to 95 percent loss without pesticides.

TABLE 7-2
CROP LOSSES DUE TO PESTS

Causes of Crop Loss INSECTS	1951- 1960	1942- 1951	1910- 1935	1904	Losses if No Pesticides Were Used
Losses in billions of dollars	3.8	1.9	0.6	0.4	4.8[e]
Percentage of crop lost	12.9	7.1	10.5	9.8	16.3[e]
CROP DISEASES					
Losses in billions of dollars	3.6	2.8	NA	NA	4.2[g]
Percentage of crop lost	12.2	10.5	NA	NA	14.2[g]
WEEDS					
Losses in billions of dollars	2.5	3.7	NA	NA	3.0[h]
Percentage of crop lost	8.5	13.8	NA	NA	10.2[h]
TOTAL LOSS					
Losses in billions of dollars	9.9	8.4	NA	NA	12.0
Percentage of crop lost	33.6	31.4	NA	NA	40.7
POTENTIAL PRODUCTION VALUE (billions of dollars)	29.5[i]	26.7	5.7	4.1	29.5[i]

[e]Assumes that, in addition to total insect-caused crop losses of $3.8 billion on both treated and untreated acres (1951-60 data), an additional $1.0 billion loss would occur if the 5 percent of the crop acres receiving insecticide treatment (Table 1) were left untreated. As further explanation, the $4.8 billion (16.3 percent) crop loss figure if no insecticide were used for insect control is based on the following: An overall 12.9 percent crop loss due to insects occurs on both treated (5 percent of toal U.S. agricultural acreage) and untreated (the remaining 95 percent) farmland. On the treated acres, 71 percent are planted with cotton, corn, fruit, and nuts. If all cotton were untreated, losses were assumed to average 32 percent, losses on all untreated corn assumed to average 15 percent, and losses on all untreated fruit and nuts to average 60 percent. A 12 percent loss was assumed for all the other untreated acres. The estimated value of cotton was $2.5 billion, corn $4.4 billion, fruit and nuts $1.4 billion, and all others $21.2 billion. It follows that crop losses due to insects if

no insecticides were used would be: 32 percent of ($2.5 x 10⁹) +
15 percent of ($4.4 x 10⁹) + 60 percent of ($1.4 x 10⁹) + 12
percent of ($21.2 x 10⁹) = $4.8 billion.

*g*Assumes that, in addition to total crop losses of $3.6 billion on
both treated and untreated acres due to crop disease (1951-60
data), a $0.6 billion loss would occur if the 0.5 percent of the
crop acres receiving fungicide treament were left untreated. The
$4.2 billion (14.2 percent) crop loss figure if no fungicide were
used for crop disease control is based on the following: An
overall 12.2 percent crop loss due to diseases occurs on both
treated (0.5 percent) and untreated (99.5 percent) acres. On the
treated acres, 51 percent of the acres are planted with peanuts,
potatoes, citrus, and apples. Untreated peanut losses were
assumed to average 25 percent, losses on untreated potatoes
assumed to average 30 percent, losses on untreated citrus
assumed to average 60 percent, losses on apples assumed to
average 80 percent, and losses on all other crops assumed to
average 12 percent. The estimated value of peanuts was $0.18
billion, potatoes $0.61 billion, citrus $0.46 billion, apples $0.25
billion, and all other $28.0 billion. Thus crop losses due to
diseases without fungicides are: 25 percent of ($.18 x 10⁹) + 30
percent of ($0.61 x 10⁹) + 60 percent of ($0.46 x 10⁹) + 80
percent of ($0.25 x 10⁹) + 12 percent of ($28 x 10⁹) = $4.2
billion.

*h*Assumes that, in addition to the $2.5 billion loss (1951-60 data)
due to weeds, the 12 percent of the acres receiving herbicides
would require $0.5 billion in cultivation and other weed control
practices to provide equally effective crop production. The $3.0
billion (10.2 percent) loss figure due to weeds if no herbicides
were used is based on the following: A $2.5 billion loss due to
weeds occurs on the 890.8 million acres of treated and untreated
crop lands (pastures included). On the treated acres, 46 percent
are corn, wheat, sorghum, rice, and pasture. If substitute
practices of weed control (cultivation and other practices) were
employed for herbicides, the additional cost per acre is estimated at $5 for corn, $5 for wheat, $3 for sorghum, $10 for rice,
$5 for pasture, and $5 for others. The millions of acres treated
with herbicides are corn 37.8, wheat 15.3, sorghum 0.9, rice 1.0,
pasture 5.4, and other 39.3. Thus crop losses due to weeds which
include the alternative control costs are: $2.5 x 10⁹ + $5
(37.8 x 10⁶)+$3(4.9 x 10⁶)+$10(1.0 x 10⁶)+$5(5.4 x 10⁶)+
$5(39.3 x 10⁶) = $3.0 billion.

*i*Pest losses (for 1960) + Actual Crop Production (for 1960) +
Potential Crop Production, or $9.9 billion + $19.6 billion =
$29.5 billion.

Organic Farming

Using the figure of $2.1 billion to represent the additional loss incurred by eliminating pesticides, an estimate can be made of the dollar return per dollar invested in pesticides for crop protection. With about $0.56 billion spent in 1966 for pesticides in agriculture and assuming application costs for labor and machinery to be one-third the cost of the pesticide materials (or about $0.186 billion), the return per dollar invested for pesticide control is about $2.82. (That is, $0.56 billion plus $0.186 billion equals $0.746 billion; $2.1 billion divided by $0.746 billion gives us the $2.82 figure.) This estimate is somewhat below previous estimates of $4 to $5 returns, but the latter are based on different methods of calculation.

An estimate of the increase in retail food prices due to the additional 7.1 percent loss of agricultural productivity can be projected. Because farm products have low-level elasticity (meaning that a small change in commodity supply causes a significantly larger change in price), for every 1 percent decrease in quantity of farm products produced there is roughly a corresponding 4 percent increase in price; therefore, the 7.1 percent increased loss would result in a 28.4 percent increase in farm product value. This would amount to about a 9 percent increase in retail food prices. If prices did rise, farmers probably would respond with efforts to increase output of the affected crops to establish a new quantity and new equilibrium price. Hence, through farmers' efforts to plant more, the 7 percent crop loss gradually would be reduced along with prices.

For a few crops like apples, an increase in acres planted would not be a practical substitute for pesticide use, because codling moth larvae and apple maggots inside the apples make this fruit unsalable. Oranges also are in the same category as apples because of the currently high "cosmetic standards" now expected by the consumer.

This leaves us with the question as to whether there would be undesirable consequences for our food supply if elimination of pesticides caused an increase of 7.1 percent in crop losses. As a matter of fact, this nation normally produces an estimated surplus of 10 percent annually. Thus, if no pesticides were used, the supply of food for the nation would be ample, even without the compensating measures described above. However, quantities of certain fruits and vegetables such as apples, peaches, plums, oranges, onions, potatoes, and cabbages would be significantly reduced. Because of this, we might have to use substitutes for some of the fruits and vegetables we normally like to eat.

The loss in some fruits and vegetables would not be quite as large as the estimate if "cosmetic standards" were modified. Although safe and nutritionally sound, some fruits and vegetables are not sold in the market today because of their less-than-perfect outer appearance. For example, oranges with dark blemishes or scales on the peel are not sold, but these skin blemishes do not adversely affect the fruit. Also, cabbages with insect-caused holes in the outer leaves are not sold, but with these outer leaves removed, the cabbages are perfectly wholesome. The public could be educated to be more concerned about pesticide contamination of fruits and vegetables than about their outward appearance.

One possible alternative to pesticides already mentioned involves increasing the number of crop acres planted to offset pest losses. This would mean reactivating some of the nearly 60 million acres now diverted from crop production. This alternative would be satisfactory for crops such as corn and cotton, but as mentioned before, it would not work for crops such as apples and peaches.

A second alternative is to plant some major crops in geographic regions where pests are generally less numerous, thus decreasing use of pesticides. Implementing such a change would be sociologically difficult, but the advantages obtained warrant further consideration. The importance of geographic regions for production is well illustrated by the codling moth pest in apples. In the South, there are three complete generations of the pest per year, whereas in the far North there may be only a single generation. Obviously, less insecticide is needed in the northern region for control of the codling moth. Also, it should be pointed out that there are varieties of apples which are significantly resistant to the codling moth. Implementing sanitation and other cultural controls would also contribute to reducing this pest.

Pest problems with vegetables also vary in severity in different climatic regions. For example, in the Southeast, 100 percent of the potato acreage received insecticidal treatments during 1966, whereas in the northern plains, only 42 percent received treatments. Similar differences occurred with fungicides. In the northern plains, 94 percent of the potato acres were treated with fungicides; in the mountain region, only 19 percent of the potato acres were so treated. Although other factors are involved, these figures suggest that some regions have fewer pest problems and in that respect may be superior to other regions for cultivation of a particular crop.

Another alternative is to rotate crops. For many decades crop rotation proved to be an effective and profitable practice. Unfortunately, in the past few decades, pesticides have been substituted for this practice in the case of some crops. In other cases, employing one pesticide may prevent the rotation of a crop. For example, some herbicides used in cultivating corn may hinder the capacity to rotate that plant with oats, soybeans, and other plants. This is because the herbicide residues may kill the latter, susceptible plants. As a result, in some cases there is a tendency to grow corn crop after corn crop on the same land, and this may increase insect, disease, and weed pest problems. For instance, corn rootworm is a pest problem which may follow repeated corn croppings, requiring application of additional insecticides as a countermeasure. This contributes to pollution and also increases the overall costs of pest control.

Another long-time, effective control technique for some pests is so-called crop sanitation, which means destroying crop remains after harvest. This practice eliminates a large portion of the corn borer population. In the 1930s and 40s, fall plowing of corn stubble and stalks was widely used as a control measure for the corn borer; however, this measure was never economically evaluated.

Resistant varieties of crop plants could replace some of the more susceptible varieties. Although only a few insect-resistant varieties are presently available, these have been highly effective in reducing crop damage. For example, the Hessian fly, a serious pest of wheat, is well controlled by planting resistant wheat varieties. The reproduction of chinch bugs differs greatly on two varieties of sorghum. On one variety (dwarf yellow milo), the bug produces an average of 99.4 offspring, whereas on a highly resistant variety (Kansas orange sorgo), the production of offspring averages only 0.3.

Sometimes years of research are required to find resistant factors in plants and then to incorporate these factors into a plant variety which has all the desirable yield and quality characteristics. Although this process usually is assumed to be a long-range undertaking, resistant varieties have been developed in three to five years.

In addition to those bioenvironmental controls

which could be implemented immediately for control of certain pests, research should be undertaken to develop new bioenvironmental controls. Potential exists in controls obtained by use of parasites (including pathogens), predators, attractants (chemical and physical), sterile males, and genetic manipulation.

Insecticide usage, for instance, could be drastically cut if bioenvironmental controls were developed for just a few major crop pests. For example, an estimated 40 percent of all insecticide applied annually in the U.S. is employed against only three pests: the cotton boll weevil, the cotton bollworm, and the apple codling moth. Development of effective bioenvironmental controls of these three pests would significantly reduce total insecticide use.

Pesticides are valuable pest-management tools, and this deserves reemphasis. The prime difficulty lies in man's tendency to substitute pesticides for effective bioenvironmental controls. A number of agriculturalists believe that pesticides should be employed primarily as "stop-gap" or "fire-fighting" tools, and sound bioenvironmental controls should be relied upon as the primary control method. They favor a "systems" approach that takes into consideration the entire complex of agricultural, economic, public health, and environmental factors. Such an approach is commonly referred to as integrated pest management. Immediate reductions in pesticide use would be possible by substituting "treat-when-necessary schedules," based on the actual measurements of pest populations, for the currently employed "routine spray schedules" which waste pesticides, contribute to pollution, and increase food costs. Estimates suggest that farmers could reduce insecticide use 35 to 50 percent with little or no effect on crop production by this "treating-when-necessary" approach.

A further reduction in pesticide use would also be possible if the current policy favoring 100 percent pest destruction were replaced by lower-level control based on sound economic-threshold densities.

In conclusion, pesticide use in the U.S. could be reduced significantly if: 1) bioenvironmental pest controls, which were replaced by reliance on pesticides, were again put into full practice wherever possible; 2) some or all of the 60 million acres currently diverted at a cost of $3 to $4 billion annually were planted to help balance the increased crop loss resulting from a reduction in pesticide use; 3) a "treat-when-necessary" program based on monitoring pest populations were initiated and aircraft spray drift were counteracted; and 4) the public was educated to be concerned for the safety of their fruits, vegetables, and other produce and to attach less importance to "cosmetic appearance."

David Pimentel
Professor, Department of Entomology and Section of Ecology and Systematics, New York State College of Agriculture and Life Sciences, Cornell University.
Reprinted from Environment, *vol. 15, No. 2, copyright © 1973 Scientists' Institute For Public Information.*

Biological and Cultural Practices for Insect Control

A combination of biological and cultural practices can be utilized to reduce risk of insect losses. They are the oldest methods of insect prevention, and some so new that they probably will not be in a usable stage for many years.

Crop rotation: This is one of the oldest. Some recommendations for rootworm control date back to the 1870s in Missouri. Rotations have been recommended for decades for specific insects. Crop rotation is not effective for insects that have high mobility. It is limited to insects that stay (from one year to the next) in the location where the crop is grown as in rootworm reduction in continuous corn production. It does not work for greenbugs, European corn borers, or alfalfa weevils because they are mobile.

The first recommendation for preventing rootworms in corn is rotation. If you plant anything else, rootworms normally will not bother that crop. Corn following other crops should be free of serious rootworm infestations 80 percent of the time — but not 100 percent.

Trap crops: Another cultural approach that has been under research for several years is utilization of a trap crop. A trap crop is planting corn late, or using several varieties that mature over a longer period of time to attract beetles. The extended period of silk and pollen production continues to attract beetles to feed and deposit eggs. After eggs are laid, cut the crop for silage, and do not plant corn there next year. No one knows yet how many acres would need to be in trap crops. Perhaps in a few years, this may be a practical cultural practice.

Planting and harvesting: Manipulation of planting or harvesting times will not control insects, but can reduce the loss to insects. Altered planting times have been used for a long time, and many are still used today. Hessian fly and the date of planting wheat is an example. We still recommend, as far as insect control is concerned, to plant after fly-safe dates.

Another example is alfalfa weevil control. If alfalfa is at a certain stage of growth when weevils are beginning to chew it up, we recommend the alfalfa be cut, rather than using chemicals. Usually if alfalfa is in bud or early bloom stage when weevils are acting up, put the weevils in the hay too, they may add to the protein. Also, by cutting early, the larvae fall onto the bare ground and are subject to perishing due to exposure and birds.

European corn borer infestations frequently are severe in the fall, due to the usual increase of the second brood. The second brood is difficult to control with chemicals as moths deposit eggs over a long period of time. It is better to harvest early to prevent losses rather than to attempt chemical control, with the exception of seed-production fields.

Weed control: We do not think of weed control to reduce insects, but it can be important in a few situations. Weed pollen is an alternate food for rootworm beetles. Weed control is very important in crop rotation to prevent rootworm injury in corn. Rootworm adults will move to weeds that are flowering in August in soybeans, stubble, or other fields. After feeding on weed pollen, beetles will deposit eggs in the locations they are feeding. The next year there may be enough eggs to cause havoc to corn.

Resistant varieties: Probably the most useful and practical of cultural practices is using resistant varieties as they are developed. The term "resistance" may be misinterpreted as immunity. Resistant varieties are not usually immune to specific insects, so are subject to injury. Resistance may be described as *tolerance* in a plant causing the plant to react differently to insect feeding. A tolerant variety will grow more vigorously under the same insect population pressure than a nontolerant variety. In some cases, tolerance may not be enough to be a sole source of yield protection, but is helpful. An example of some of the alfalfa varieties tolerant to weevil are Team, Weevilchek and Gladiator. These

267

varieties are attacked by weevils, but regrow more rapidly than others. Many varieties of corn are more tolerant, especially during early growth, to European corn borers—but they tend to lose tolerance as they grow older and have little or no tolerance to the second brood. Another example in corn is root regrowth capability. Some varieties appear to grow roots faster than others—an important character when rootworms are present.

One of the earlier uses of resistance in crop productions is Hessian fly "resistant" wheat. Some varieties are more tolerant, but do not have immunity. Also, insects may change to adapt to resistant varieties—as the Hessian fly has.

Another factor of resistance has a big name—"antibiosis;" it simply means against life. The chemistry of some plants will cause lower reproduction rates of insects that feed on them, smaller insects, and perhaps in extreme cases, inability of insects to survive to reproduction stages. Another resistance factor is preference. Some insects may not like certain varieties as well as others.

Adaptability of crop: A crop better adapted to your growing conditions would tolerate more insect injury than those that are poorly adapted. A well-adapted crop will produce more than one that struggles to survive because of simple environmental conditions . . . even with the insects.

Growth stimulation: By efficient and accurate fertilization and water management, we can reduce insect losses. The insects are not controlled, but growth under the most ideal conditions that can be provided helps plants survive and produce in spite of insects.

New crops: Sometime in the future, we may have to grow new crops to defeat insect enemies. In Enterprise, Alabama, there is a rather elaborate monument to the boll weevil. Many years ago the boll weevil destroyed so much cotton, it was not feasible to produce it as a crop. There were no chemicals, or cultural practices that worked. They planted peanuts. Peanuts were more profitable than cotton. The monument was erected to the boll weevil as a thanks to the insect that forced farmers out of cotton production and into a more profitable crop.

Robert E. Roselle
From a talk at the "Organic Residues and By-Products In Crop and Animal Production Workshop," held by the University of Nebraska, Dec., 1975.

Herbicide 2, 4-D Increases Insect and Pathogen Pests on Corn

Since 1945, increased losses due to attack of insects and pathogens have been reported for crops in spite of greater efforts in pest control. How much, if any, of this increased loss caused by pests is due to the ecological and biochemical impact of herbicides on crops is unknown, but in a number of instances herbicides have been reported to increase pest problems on various crops. Our laboratory and field tests were designed to determine what influence the use of 2, 4-dichlorophenoxyacetic acid (2, 4-D) has on the susceptibility of grain corn to the European corn borer, corn leaf aphid, and the southern corn leaf blight.

In 1973 a preliminary study was made of the impact of 2, 4-D herbicide on corn leaf aphid and European corn borer populations in the corn variety Pennsylvania 290. The three treatments of 2,4-D per hectare were: 1) untreated (control); 2) 0.14 kg; and 3) 0.55 kg. (normal use). The herbicide spray was directed at the base of knee-high corn plants and toward any weeds, and all plots were cultivated for weed control. Aphid counts were made on sixty ears of corn selected systematically from each of these plots during late September. The number of aphids, following the three treatments, were: 1) 618; 2) 1,388; and 3) 1,679. Corn borer infestations were measured in late August, and the percentages of plants in these plots that were infested with corn borer larvae were 16 percent after 1); 24 after treatment 2); and 28 after treatment 3).

More extensive field tests were made in 1974 on three-row plots (seventy to ninety plants) 24 by 7 m in size. Four treatments: 1) untreated (control); 2) 0.14 kg of 2,4-D per hectare; 3) 0.55 kg of 2,4-D per hectare (normal use dosage); and 4) 4.4 kg of 2,4-D per hectare were used; techniques were the same as in the 1973 tests. Aphid counts made on the tassels of the corn were significantly higher in the plots treated with 0.14 and 0.55 kg of 2,4-D per hectare than in the untreated plots. These numbers were for 1) 1,420; 2) 2,449; 3) 3,116; and 4) 2,023. The percentage of corn plants attacked by the corn borer were 63 percent after 1); 83 after treatment 2); 70 after treatment 3); and 63 after treatment 4).

In laboratory tests the single hybrid OH 51A × B8 corn was grown in a growth chamber at temperatures of 28° to 29°C. After four weeks (when the corn was forty to fifty cm tall) ninety ml of 2,4-D solution was applied to the soil in each pot at concentrations of 0, 5, 20, 80, and 320 parts per million (ppm). The 20-ppm concentration approximated the field dosage of 2,4-D of 0.55 kg/ha. Then two weeks after herbicide treatment each experimental corn plant was infected with five first-instar corn borer larvae placed in the whorl of the plant. Mean weights of corn borer pupae obtained from larvae reared on corn treated with 5, 20, and 80 ppm of 2,4-D were significantly heavier than larvae reared on untreated corn. Moths reared on plants receiving 5, 20, and 80 ppm of 2,4-D on the average produced more egg masses per female than those reared on untreated plants.

Corn plants from the corn borer experiments were tested for total protein. Corn plants receiving 5, 20, and 80 ppm of 2,4-D contained higher levels of protein than the untreated plants and plants receiving 320 ppm of 2,4-D. The increased protein of the treated plants probably improved the nutrient content for the corn borers, corn leaf aphids, and the southern corn leaf blight pathogens.

The impact of 2,4-D on the susceptibility of corn to southern corn leaf blight was studied in other tests. Seven days after planting, the corn was treated with 2,4-D at 10, 20, 40, 100, and 200 ppm, and one group of plants was left untreated. After six days the plants were spray-inoculated with 100 ml of a spore suspension standardized to 11,500 spores per milliliter. Pathogen infection was determined by counting all lesions larger than one cm in length. Corn plants treated with 20, 40, 100, and 200 ppm of 2,4-D had significantly more lesions (greater than one cm in length) than the untreated plants and the plants with the lowest dosage of 2,4-D (10 ppm).

The results of this investigation demonstrate that increased risks of attack by insects and disease on corn may result from herbicide treatments. Additional studies are needed on other crop plants on which herbicides are used to determine the potential impact herbicides are having on plant protection programs.

I. N. Oka
Department of Pests and Diseases of the Central Re-

Organic Farming

search Institute for Agriculture, Department of Agriculture, Bogor, Indonesia

David Pimentel
Department of Entomology and Section of Ecology and Systematics, Cornell University
Reprinted from Science magazine, July 1976.

A New Insect Control Method

Frank Batey grows peanuts and soybeans in bug-ridden Archer, Florida. For years he and his father used guthion, toxaphene, sevin, methyl parathion, and other pesticides to combat the fierce semitropical insects that would otherwise have ravaged his crops. Then Mike Sipe, a pest-control specialist from the Gainesville area, suggested a simple method of dealing with the bugs. Today Batey's fields are crawling with insects, but he sustains no important damage to his crops. So far he's saved $5,000 in insecticide costs. Best of all, he says, "I don't have to breathe that peanut poison anymore."

The method is this: Batey goes into his fields and collects a cup or two of the insects that are damaging his crops. He puts them in the blender with some water, strains the result, dilutes it, and sprays it on his fields. So effective has the method been over two years that in 1976, his third year, he didn't even have to spray the bug juice on his fields.

The discovery of this method—which sounds too good to be true—began with a short article in the December 1972, issue of OGF, entitled "Control the Cabbage Looper." It described the way Francis R. Lawson, a former USDA entomologist, controlled this major pest. He searched his fields until he found dead, dying, or diseased loopers, which, he reasoned, were infected with viruses that attack loopers specifically.

These diseased insects were then put through a blender, diluted, and sprayed onto the crops, infecting all the healthy loopers. Sipe, reading Dr. Lawson's manuscript, advised him to encourage *OGF*'s readers to try it. Dr. Lawson complied. A year later, we received a letter from Eleanore M. Bubb of Phoenix, Arizona, who said that she had tried it on the western grape skeletonizer. "I caught four adult moths and scraped eggs from three leaves," said the letter. "I put them into my blender with a quart of water and liquified them. My two grapevines were sprayed with this concoction the same day. I have seen no adults since then. The egg clusters laid on the leaves hatched, but the insects died after making only a small spot on the leaves. After one heavy rain I noticed some of the leaves had been skeletonized, so I used the solution again. I have seen no adults or caterpillars since," she said.

Sipe was immediately struck by the fact that she had used *healthy* moths—or had not reported that they were diseased. Mike then tried it with excellent results on his own garden and suggested to Frank Batey, who was then taking over the farm operation from his father and was employing Sipe to reduce pesticide costs by recommending integrated controls, that he try the method on the voracious peanut and bean pests.

"We had cabbage looper, stink bug, army worm, velvet bean caterpillar, granular cutworm, southern corn borer, and other pests on our crops," Batey says. "I collected samples of them all, except for the southern corn borer, which makes a webbed tunnel underground and attacks the pegs of the peanuts. It's just too difficult to find enough of them to make a spray. But on the others, the method did an outstanding job.

Batey tells of the farmer who lives near him. "Last year he sprayed so hard that his peanuts almost died. He came over and looked at our crop. There were ants, earwigs, and beetles crawling around on the ground and on the plants, but none of the major pests. He didn't know what to make of it." He says his neighbors are "hard-nosed about new methods," and he hasn't told them about the method yet. "They would think it's hogwash. But I can prove it works. I *know* it works."

Sipe speculates on why it works:

"There may be three possible reasons," he says, "and of course there may be other reasons that we just haven't thought of. The first is pathogen activation, which is achieved by picking up a fairly large sample of insects. You're bound to get a few that harbor disease pathogens (such as viruses, bacteria, fungi, and others), even if the symptoms aren't apparent. Or it may be that healthy insects have these pathogens in their bodies—just as healthy people can harbor many diseases—but they are dormant and the insects don't get sick. This may be because the pathogens are isolated as spores, for instance, in some part of the insect and won't become active until something happens to activate them. Anyway, by putting them into a blender, you're making a solution that contains these pathogens. When sprayed over a field, they are somehow activated and start an epidemic in the insect populations that are susceptible.

"Another reason may be that the odor of ground-up bugs attracts predators or parasites," Sipe says. Batey notes that when a field is infested with worms, there's a green, rank smell. He's noticed the same smell after applying the bug juice and thinks that this may attract the enemies of the pest.

"The third reason why this may be working is that the insects' distress pheromones (substances that trigger specific behavior in insects) are released when the bugs are liquefied in the blender. These pheromones, when sprayed over a field, act as a repellent to drive away the pests. Or it may be a combination of these reasons," Sipe says.

It seems most likely that the method spreads a disease organism such as a virus, if the experience of Bill Matthews is an indication. While Batey farms a seventy-four-acre peanut allotment on his own land, he also share-crops neighbor Bill Matthews' thirty-nine-acre allotment nearby. "I let Bill know that we were going to use the bug juice to combat insects on his allotment, and he thought I was crazy. When I put the material on the crops, he thought it was hilarious. But within three or four days most of the worms were hanging dead from the leaves or lying dead in the rows. Bill had the worst insect problem I'd ever seen," Batey reports, "but the method took care of it."

"I find that a half-pound of worms will give me enough solution to treat my seventy-four acres," Batey says. He mixes five cubic centimeters of the blender solution with twenty-five gallons of water and applies it along with his fungicide applications. Four ounces of the blender liquid is enough to make about 1,500 gallons of spray when diluted. Batey also farms 100 acres of soybeans and uses the spray effectively on that crop.

Batey's peanut production was 5,251 to 5,351 pounds per acre in 1974, a tremendous harvest that won him the County Peanut Production Award. Inconsequential pest damage had a part in that record. Average production in Alachua County is between 2,000 and 3,000 pounds per acre, he says. His pesticide costs had run close to $2,000 per year, so over the three years that he's been using the bug-juice method, he's saved more than $5,000.

"The economics don't thrill me as much as the

ecology part of it," Batey says. Sipe says that Batey really got into his fields and looked and looked, then looked some more at what was happening. He studied his crop ecology until he knew which insects were pests and which were helpful bugs, and could see how they interacted. "People wishing to use this method should try to learn as much as possible about the intricate complexities of the interrelationships between plants and bugs of all kinds," Sipe says.

Organic Farmers go to Court over Spraying

What if you are an organic farmer, and your neighbor's pesticide application gets on your crop—not so that it's physically damaged, but enough to keep the produce from being certified as organically grown? Do you have a valid claim?

That was the question a judge and twelve-member jury wrestled with in Yakima, Washington, early in October. Their decision: *the organic farmer should be paid for having his crop disqualified.*

In a major legal victory already reverberating nationwide, the court awarded Patrick and Dorothy Langan of Toppenish, Washington, $5,500 in damages because an aerial application of Thiodan, used to kill Colorado potato beetles, drifted onto their land from the crop-dusting done on adjoining property. As a result, the Langans' crop of beans and tomatoes could not meet organic standards and had to be de-certified. The claim for damages was based on the rules of the Northwest Organic Food Producers' Association, which requires "decertification of crops and fields once they are sprayed with a pesticide."

The significant point is that the court decision (an eleven-to one-jury tally) established some new rules. It recognized the fact that the marketability of organically grown food was destroyed by the helicopter-spraying incident, even though no direct proof of physical injury to the food itself was evinced. After the 1973 spraying, Langan had his crop tested for pesticide contamination. It showed 1.4 parts per million of Thiodan; and while FDA tolerances allow for 2.0 ppm, organic certification requires .2 or less.

Following five days of testimony, Judge Blaine Hopp Jr. instructed the jury that it was "to return a verdict in favor of the Langans if the jury found that chemicals fell on their crop." Explains Attorney Douglas D. Peters of Selah, Washington, "The jury was not to consider who was negligent or who was at fault, but simply the event of chemicals on their property. Thus, both the seller of the chemical, the landowner, and the spray applicator were all in the same boat in the eyes of the court. The legal doctrine used by the court was 'strict liability', which means liability without fault."

Although four states have previously held spray damage "strictly liable," adds Peters, "to my knowledge this is the first case in the nation involving the claim of an organic farmer against a crop sprayer." The successful lawyer stresses that the court's decision indicates "crop spraying is a hazardous, dangerous activity. Therefore, whether careless or not, you are responsible if you damage the property of your neighbor."

Pat and Dorothy Langan, who started their farm in 1970, were the prime organizers of the Northwest Organic Food Producers' Association, a group of growers, distributors, and retailers seeking to ex-

pand marketing organically grown crops collec-
tively and confidently in the area. They drew up firm
rules for soil care, lab tests, and field inspections for
members—and they worked diligently to help the
association grow over the next several years.

Then disaster hit. They had to withdraw their own
main crop because of the spray drift. Since then, it's
been a long, hard fight to win a legal battle that
protects the organic grower.

The spray suit decision came fast and emphati-
cally. On the positive side, the Langans have been
contacted and applauded by a steady stream of
people, including other organic growers, natural
food shops, and distributors who've already bought
up their next crops. "People just living here in the
country—not farmers," Dorothy exclaims, "have
been telling us they're thrilled about it, because
they've made complaints about sprays in the past
and have been told, 'If you don't like it, move to the
city'. These folks say they like country life—except
during spray season. 'We'd wake up in the morning
with the house full of poison dust, and there was
nothing we could do about it!' "

Pat's also been contacted by the USDA Experiment
Station at Prosser, whose extension specialists have
now asked about visiting the farm and talking about
their successful organic methods. Five of the state's
organized beekeepers have expressed their delight,
too—because the case opens the door for them to
take legal action. Russell Elliott, a friend of the Langans and another NOFPA farmer, was dusted with
Sevin this past season, and plans to go to court.
"He's got a strong case," says Dorothy, "including
witnesses."

Lawyer Doug Peters has had plenty of reaction,
too: calls from several attorneys wanting copies of
the brief for similar cases; a call from the state
assistant attorney general indicating that he has

warned spray applicators and chemical companies of the implications in this decision; and even invitations to speak on the case at various meetings.

More important is the probability of an appeal. Peters says that it is most likely to be made within the thirty-day limit by the crop spray applicators—both the state and national organizations combining to combat a legal ruling which literally has them "bugged." The chemical companies have chimed in with loud laments about insurance costs and the effects on business and agriculture.

An appeal would be desirable, Peters explains, if the decision is upheld. "While spray-damage cases have been handed down and sustained in other states, the significance in this case of the lost market value to the organic grower because of the loss of certification has not been tested." An appeal, he says, would not mean a retrial, but simply a review of the evidence by the Washington State Court of Appeals, à la a U.S. Supreme Court procedure.

"I believe we'd win on an appeal," states Peters. "Then the decision would be picked up at the national level. It then becomes a reported case which attorneys in other states can cite. If it is not appealed and sustained, it remains a local ruling and has no binding effect."

"We are going to come out on the short end on the award," says Dorothy, "as it was a costly trial. In fact, our attorney didn't receive as much as he deserved for all he put into it. However, we did receive what we went after. Hopefully, a lot of people will be helped by this decision—which is truly raising havoc here in this huge agricultural valley."

M. C. Goldman

Chapter 8
Livestock

Grass Feeding for Better Meat and More Profits

With an abundance of sun, water, land, and cattle, the South has the potential to economically produce lean, tender, grass-fed beef on its high-quality forages. Compared to humans and some other mammals, beef cattle are not good converters of grain to meat. However, beef cattle are unique in their ability to convert forages into palatable red meat at a low cost-per-unit of carcass weight. Therefore, a study was conducted by P. R. Utley, R. E. Hellwig, and W. C. McCormick, at the Coast Plain Experiment Station in Tifton, Georgia, to determine the economic feasibility of finishing steer calves and yearlings to market weights on an all-forage diet. The results of their experiments reported here originally appeared in the *Journal of Animal Science*.

A total of 68 crossbred steers (32 calves and 36 yearlings) were fed either a high-energy diet or an all-forage diet in a series of trials during a four-year period. Eight calves received each dietary treatment during each year for the first two years, and nine yearlings received each dietary treatment during each year for the third and fourth years. Steers that received the high-energy treatment were full-fed a diet composed of 72.8 percent ground shelled corn, 20 percent peanut hulls, and 7.2 percent protein supplement. The high-energy diet was fed to calves for 196 days and to yearlings for 161 days. Steers that received the all-forage diet were fed Bermuda grass pellets free choice in drylot for about 56 days, transferred to small-grain pasture (oats, slender, aerial part, fresh, or rye, aerial part, fresh, for about 120 days, then returned to drylot and fed Bermuda grass pellets until slaughter weights were reached. Calves in the all-forage groups were fed and grazed for 240 days and yearlings were fed and grazed for 205 days.

Individual unshrunk weights were obtained initially, at twenty-eight-day intervals during the trials, when dietary changes were made, and at the end of each trial. Total feed consumption (excluding pasture) by each lot was calculated at times corresponding to animal weigh-days. Subsequent to the termination of each trial, the steers were slaughtered at a local packing company and carcass data were obtained. Hot carcass weights were obtained on all cattle and were used to calculate chilled carcass weights by employing a standard 2 percent cooler shrink. Dressing percentage was calculated using slaughter weight and chilled carcass weight. Twenty-four hours after slaughter, the left half of each carcass was separated between the twelfth and thirteenth ribs and scored by an official USDA grader to the nearest one-third of a grade. Simultaneously, the carcasses were also scored as to yield-grade and marbling. Carcass fat thickness was measured at the twelfth rib separation at two-thirds the distance from the distal end of the longissimus muscle. Results are shown in Table 8-1.

Each year, winter pastures were planted as early in October as the prevailing season would permit. Oats (*Avena sativa*) were planted the first and third years whereas Abruzzi rye (*Secale cereale*) was planted the second and fourth years. Seedings were made in prepared seedbeds using conventional planting practices. When the small-grain plots were about six to eight inches tall and well rooted, the steers assigned to the all-forage treatment were put on pasture and permitted to graze continuously. Steer calves were stocked at one steer per acre and yearling steers were stocked at one steer per 1.3 acres. The grazing periods were continued until the winter pastures were totally utilized from the standpoint of producing good gains. In all years, some residue that would have been of value for maintaining breeding or stocker cattle was left in the field. The crude protein content of the oat and rye pastures ranged from 20 percent at the start of the grazing period to 11 percent at the end of the grazing period.

Average feeder-steer prices of $45 per 100 pounds for calves and $43 per 100 pounds for yearlings were

TABLE 8-1
FEEDLOT PERFORMANCE AND CARCASS CHARACTERISTICS OF CALVES AND YEARLINGS FINISHED ON A HIGH-ENERGY DIET AND AN ALL-FORAGE DIET

Item	Calves		Yearlings	
	High-energy	All-forage	High-energy	All-forage
No. of steers	16	16	18	18
Avg initial wt, lb (kg)	488(222)[a]	492(224)[a]	670(305)[b]	672(306)[b]
Avg final wt, lb (kg)	1,042(474)[a]	1,053(479)[a]	1,165(530)[b]	1,147(521)[b]
Avg daily gain, lb (kg)	2.84(1.29)[a]	2.33(1.06)[b]	3.07(1.40)[a]	2.31(1.05)[b]
Total feed consumed per steer, lb (kg)				
Ground shelled corn, lb (kg)	4,020(1,827)	—	4,111(1,869)	—
Peanut hulls, lb (kg)	2,894(1,315)	—	2,993(1,360)	—
Supplement, lb (kg)	804(365)	—	823(374)	—
Bermudagrass pellets, lb (kg)	—	1,497(680)	—	2,203(1,001)
Small grain grazing, acres (ha)	—	1.0(.41)	—	1.3(.53)
Avg carcass data				
Chilled carcass wt, lb (kg)	608(276)[a]	558(254)[b]	672(305)[c]	630(286)[d]
Dressing percentage	59.84[a]	54.32[b]	59.50[a]	56.69[b]
Marbling score[e]	4.4[a]	3.5[b]	4.2[a]	3.9[b]
Yield grade[f]	2.8[a]	2.1[b]	2.6[a]	2.4[b]
Carcass grade[g]	11.2	9.4	10.6	10.2
Fat over rib, inches (cm)	.45(1.14)[a]	.32(.81)[b]	47(1.19)[a]	40(1.02)[b]

[a,b,c,d] Means on the same line bearing different superscript letters are different (P<.05).

[e] Marbling score: 3, traces; 4, slight; 5, small.

[f] Yield grade as specified by U.S.D.A. where values range from 1 to 5 with values nearer 1 indicating carcasses with a higher percentage of retail cuts.

[g] Carcass grade: 9, low good; 10, medium good; 11, high good.

used as the purchase price. Slaughter-steer prices of
$65.75 per 100 pounds of hot carcass were used to
calculate receipts for all steers. Based on the estimated cost and resource requirements for producing
small-grain pastures, a grazing charge of $60 per
acre was calculated to be the current cost and was
used in calculating the cost of steer grains. Other
feed ingredients used were priced according to current price quotations.

Body-weight gains for calves fed the two diets
were similar for the first 28 days of feeding, different
between 28 and 56 days, and then remained almost
parallel for the remainder of the feeding period.
Body-weight gains for the yearling steers were
similar for the first 56 days, were divergent from 56
to 112 days, and then parallel for the remainder of
the study.

Animals fed the high-energy diet gained 2.97
pounds daily and steers fed the all-forage diet gained
2.33 pounds daily. Body-weight gains for calves and
yearlings were almost identical when fed the all-
forage diet. However, average gain for yearlings was
8 percent faster than for calves when fed the high-
energy diet.

Calves fed the high-energy diet consumed an
average of 20.5 pounds of the grain mixture per day
compared with 25.5 pounds per day for the yearling
steers. As a result, the calves required about 7.2
pounds of feed per pound of gain and yearling steers
required about 8.3 pounds of the high-energy diet
per pound of gain. These results are consistent with
calculated maintenance and gain requirements for
calves and yearlings. Steer calves fed the all-forage
diet required a total of 1,497 pounds of Bermuda
grass pellets and one acre of small-grain grazing to
produce 561 pounds of body-weight gain. The yearling steers fed the all-forage diet were less efficient
and required an average of 2,203 pounds of Bermuda

grass pellets plus 1.3 acres of grazing to produce 475 pounds of body-weight gain.

Steers fed the high-energy diet dressed out about 7.5 percent more of their live weight as carcass when compared with all-forage-fed steers. Therefore, the difference between actual carcass gains for steers on the two dietary treatments would be greater than the observed differences in live-weight gains due to differences in body fill.

Differences between other carcass characteristics were similar but more pronounced than previous research showed (1971). All-forage-fed steers had less marbling, lower yield grades, and less fat covering over the rib eye than steers fed the high-energy diet. However, carcasses from steers in all groups averaged grading from medium to high good. With the exception of chilled carcass weights, no difference was detected in carcass characteristics between calves and yearlings.

An economic analysis of the two finishing systems using either steer calves or yearlings is presented in Table 8-2. The actual prices received per 100 pounds of carcass for steers on the two diets were almost identical during each of the four years of the study. Therefore, an average price of $65.75 per 100 pounds of hot carcass weight was used to calculate the sale value of all steers. Total feed costs for producing slaughter steers ranged from $93.63 to $181.22 per steer. When the high-energy diet was fed, the feed cost per 100 pounds of gain was $32.22 for calves and $36.61 for yearlings. With the all-forage diet, the feed cost per 100 pounds of gain was $16.69 for calves and $23.07 for yearlings. Returns to capital, land, labor, and management were $9.66 and - $18.63 per head, respectively, for calves and yearlings fed the high-energy diet, compared with $59.19 and $23.97 per head, respectively, for calves and yearlings fed the all-forage diet.

TABLE 8-2
ECONOMIC ANALYSIS OF FINISHING CALVES AND YEARLINGS ON A HIGH-ENERGY DIET
AND AN ALL-FORAGE DIET

Item	Calves		Yearlings	
	High-energy	All-forage	High-energy	All-forage
Receipts:				
Slaughter steers, $[a]	407.76	374.22	450.68	422.51
Expenses:				
Feeder steer, $[b]	219.60	221.40	288.10	288.96
Feed cost, $[c]	178.50	93.63	181.22	109.58
Total expenses, $	398.10	315.03	469.32	398.54
Returns to capital, land, labor, and management, $	9.66	59.19	-18.63	23.97

[a]Slaughter steers valued at $65.75 per 100 lb of hot carcass weight.
[b]Feeder steers calves valued at $45.00 per 100 lb and yearling steers at $43.00 per 100 pounds.
[c]Feed prices used: Corn, $.05/lb; supplement, $.09/lb; peanut hulls, $12/ton; bermudagrass pellets, $45/ton; small grain grazing, $60/acre.

Growth Stimulants, and Trying to Control Them

The spectre of cancer has long haunted America. More than any other disease or ailment, it looms over our heads and lurks in the corners of our imaginations. It wasn't too long ago that the word "cancer" could not be uttered without casting an uneasy pall throughout the room.

The image of cancer as a random, unpredictable killer is not entirely accurate, however. Cancer experts believe that 60 to 90 percent of all human cancers are related to environmental factors and, therefore, that most are preventable. For certain types of cancer, relationships have been clearly pinpointed, as in the case of cigarette smoking and lung cancer.

In other instances, the relationships may be more subtle. Many suspect the now-notorious synthetic hormone *diethylstilbestrol* (DES) of subjecting beef eaters to an increased risk of cancer.

Since November 1954, when it was approved for use in cattle by the FDA, DES has been the most popular growth stimulant drug among cattle raisers, and understandably so. A 10 mg DES pellet implanted behind the animal's ear would cost from twelve to fifteen cents, but would bring a beef animal to a market weight of 1,000 pounds thirty-five days sooner and with 500 pounds less grain than would be the case without a growth stimulant.

The DES advocates were faced with one nagging problem, however: many experiments, dating as far back as 1941, show that DES is a carcinogen. As one National Cancer Institute specialist put it, DES has long been "one of the chapters in a textbook" on cancer.

By 1973, even though over twenty countries had banned the use of DES in meat production, 80-85 percent of America's cattle were being treated with the drug. However, despite tightened restrictions on its use, radioactive tracer studies conducted by the Department of Agriculture found DES residues in edible beef tissue even when the FDA regulations were strictly followed. So, in 1972 and 1973, FDA finally withdrew all approvals for DES use in food animals, first for use in feed, then in pellet implant form, only to have this decision overturned by the U.S. Court of Appeals for the District of Columbia in 1974. The court ruled that the agency had not provided DES manufacturers adequate opportunity for a public hearing. Thus, on an unfortunate procedural point, DES was allowed to return to the marketplace, where it remains today.

FDA is continuing its efforts to prohibit the use of DES in beef cattle. In December 1975, the agency proposed again to ban veterinary use of DES, this time allowing DES interests an opportunity to request a hearing on the decision. By the time the deadline for submission of these requests arrived in February, 1976, four such petitions for a hearing had been submitted, all by DES manufacturers: Dawes Laboratories, Rhodia's Hess and Clark, Vineland Laboratories, and, jointly, Franklin Laboratories and Fort Dodge Laboratories, subsidiaries of American Home Products.

Even if the hearings take place and the DES ban is upheld, several disturbing questions remain unanswered, however.

While the FDA ban on DES implants was in effect, from April 1973 to January 1974, most cattle raisers did not give their animals a reprieve from growth-promotant drugs. In fact, an FDA Inflationary Impact Statement said, "Even with the ban on DES implants effective for most of 1973, more cattle slaughtered in 1973 and 1974 had been implanted with estrogens than during the preban year."

The Statement added, "It seems apparent that additional supplies of substitute implants can be produced to offset the restriction on the DES implant products. . . . We do not know any reason why this same substitution effect would not again occur with a second ban on DES."

Obviously, it is important to look into the nature of these "substitute implants," as well as whatever other drugs or devices may be used in lieu of DES. Unfortunately, the more one looks, the more apparent it becomes that banning DES will not necessarily result in safer meat.

Next to DES in popularity for use as growth promotants are two other estrogen drugs, sold under the trade names Synovex and Ralgro. Both are more expensive than DES—before the April 1973 ban, DES implants sold at twelve to fifteen cents a dose, while comparative applications of Synovex and Ralgro went for seventy-five to ninety-five cents. Nonetheless, farmers used large quantities of these drugs when DES was unavailable.

There are two forms of Synovex: Synovex S, which is restricted by Section 522.1940 of the Code of Federal Regulations (CFR) to use as a subcutaneous (under the skin) ear implant in steers only; and Synovex H, cleared by Section 522.842 of the CFR as a subcutaneous ear implant in heifers only. Synovex S contains estradiol benzoate and progesterone, and Synovex H contains estradiol benzoate and testosterone propionate, all three of which differ from DES in that they are natural, rather than synthetic, hormones.

Ralgro, which was approved by Section 522.2680 as a subcutaneous ear implant for beef cattle, contains zeranol, a by-product of corn fermentation with some estrogenic properties.

FDA notes that, as growth stimulants, Synovex's and Ralgro's "efficacy relative to DES is comparable." Studies at an agricultural experiment station at South Dakota State University bear this out: while untreated steers used as a control group gained 2.10 pounds a day, weight gain was increased by 36 mg implants of zeranol by 11.0 percent, estradiol benzoate and progesterone by 13.3 percent, and DES by 13.8 percent.

Taking the cost and efficiency of these drugs into consideration, FDA has stated that a ban on DES would cause most farmers to turn to Ralgro and Synovex, leading to an annual cost impact of about $503 million and a possible rise in retail beef prices of two to three cents a pound. However, this would still constitute a substantial savings for farmers over what costs would be if all growth stimulants were banned: in that event, according to the American National Cattlemen's Association estimate, prices could shoot up ten to fifteen cents a pound on beef.

Despite these figures, however, we must consider the greater question of how the substances in Synovex and Ralgro may affect the public health. Some startling information can be found in the "Johnson memo," an in-house FDA document written by Klemens Johnson of the Bureau of Veterinary Medicine to the BVM Director, C. D. Van Houweling, on "The Review of Veterinary Drug Residues and Drug Residue Monitoring Programs for Food Producing Animals."

The memo, written in September 1972, urged that "a review of the entire problem of drug residues and monitoring programs by both FDA and USDA be undertaken at the earliest possible date . . . so that screening methods, monitoring programs, and necessary regulatory methods can be updated and made consistent with current scientific knowledge." Johnson presented a specific "List of Drugs Used in Food Producing Animals," including one section on drugs that may occur in cattle tissue even though federal regulations prohibit residues or set a residue limit. Estradiol benzoate, progesterone, testosterone propionate, and zeranol all appear in this section.

It is important to note that FDA approved the use of Synovex and Ralgro on the condition that no residues of estradiol benzoate, progesterone, testosterone propionate, or zeranol appear in uncooked edible beef tissue. All these substances are identified on this list as "Possible carcinogens."

At this point, the water becomes muddier. Although all these hormones except zeranol occur naturally in human beings, they can become carcinogenic at higher-than-natural concentrations, either by themselves or in reaction with other elements. But, since hormone levels vary from person to person, and even from time to time in one individual, it is not possible to draw a definitive "safe level," above which these hormones pose a risk. Thus, in the case of estradiol benzoate and the other estrogens, the International Agency for Research on Cancer wrote, "The majority of experimental animal treatments with estrogens, which have resulted in carcinogenesis, have been at very high dose levels. There is inadequate information at present, however, to indicate the minimum dose requirements, and these could be much lower than those commonly employed in animal studies."

If the minimum safe dosage level is unknown, then it is the responsibility of FDA to act in the interests of the consumer and insure that that level is not approached. One important step in this direction would be to update the methodology currently used by the drug manufacturers and USDA for detecting traces of carcinogenic substances in meat tissue.

The Kefauver-Harris amendment gave the Secretary of HEW authority to approve certain scientific techniques to detect drug residues. The Code of Federal Regulations lists the currently approved detection procedures in its Volume 21: the Umberger mouse uterine weight test for residues of estradiol benzoate; the Hooker-Forbes bioassay for progesterone; the chick comb Bioassay for testosterone propionate; and the tissue analysis method used for zeranol.

All of these methods were also approved by the Secretary in 1972, when the Johnson memo called each one "inadequate as a regulatory method." The zeranol methodology was faulted because "it is sensitive only to twenty parts per billion," and the bioassay tests were criticized for "their nonspecificity, the long assay time involved, and the high cost per assay." In addition, John Spaulding, chief staff officer of USDA's Residue Evaluation and Planning Staff, has said of the bioassay techniques, "They are a good laboratory approach when all the variables are known, but when we don't even know if the animal has been treated with a growth promotant, we need a more specific, chemical-confirmation approach." Spaulding also said that, in using the Umberger procedure to detect traces of estradiol benzoate, "HEW has not approved the best laboratory method."

Unfortunately, better procedures are not always available for detecting traces of possible carcinogens in our food supply, but in some instances, such as in the case of DES, HEW continues to use detection methods that were approved years ago, despite the development of more sensitive, modern techniques. Although gas chromatography can pinpoint one-half part per billion of DES, HEW recognizes only the Umberger test, with a sensitivity of two to four parts per billion.

If this continues to be the case, and if the only way to protect the public is to remove possible carcinogens from the market until a reliable threshold of safety can be determined, the laws restricting FDA from doing this—the same laws which put DES back in cattle in 1974—should be changed. Peter Hutt, then general counsel of FDA, told the Senate Subcommittee on Health in February 1975, "It is terribly important that Congress change the law . . . so that we could eliminate from the marketplace animal and human drugs which we believe unsafe or ineffective for any reason, whether or not there is an imminent hazard, before determining whether a hearing is justified."

Synovex and Ralgro are not the only growth stimulants besides DES that are available to cattle raisers. One new product, Rumensin, approved by FDA in December 1975 and manufactured by former DES producer Eli Lilly & Co. through its subsidiary, Elanco, is expected to be widely used. Rumensin is an antibiotic, rather than a hormone, and is not believed to be carcinogenic, but an investigation of its safety is necessary nonetheless.

The possibility that two growth stimulants that would replace DES may also cause cancer should cause the public to demand stricter detection methods and perhaps even a revision of the law so that FDA can remove them from our food processing. Even if a possible carcinogen occurs naturally at low levels in our own bodies, this is no reason to be silent. Richard Lehmann of FDA said, "As long as we have the Delaney clause and as long as we define the carcinogenicity of hormones as we do now, if there is any increase of these drugs in our food, we should ban that drug."

GROWTH STIMULANTS: WHAT THE TESTS SHOW

SUBCUTANEOUS AND/OR MUSCULAR INJECTIONS

Drug(s)	DES	Estradiol Benzoate
Source	Lacassagne (1958)	Hooker & Pfeiffer (1942)
Dose	25 Micrograms (ug) twice a week for 12-16 weeks	16.6 or 50 ug in sesame oil once a week for 8 months or more
Animal	Male mice	Male mice
Result	Mammary tumors in 2 animals	Interstitial-cell tumors of testes in 10 of 24 animals
Drug(s)	Estradiol benzoate	Testosterone
Source	Lipschutz & Iglesius (1938)	Kimura & Nandi (1967)
Dose	20-80 ug in olive oil three times a week for 2-4 months	25 ug once a day for first 5 days after birth
Animal	Guinea pigs	Newborn mice
Result	22 of 24 animals developed multiple tumors in uterus, ventral surface of stomach, spleen, mesentary, surface of diaphragm, & other abdominal locations	Vaginal tumors in 7 of 9 females by age of about 71 weeks

Drug(s)	Estradiol benzoate & testosterone propionate
Source	Gardner (1946)
Dose	0.625-2.5 ug of testosterone propionate + 3.3-33.3 ug of estradiol benzoate once a week
Animal	Male & female mice
Result	17 mice developed mammary tumors, with average latent period of 61 weeks

(Source: IARC Monographs on the Evaluation of Carcinogenic Risk of Chemicals to Man, Volume 6)

Bob Doerschult
From Nutrition Action, April 1976

Small Cattle for Small Farms

Generations ago, Irish smallholders—who had to earn a living on very little land—bred cattle which were small, hardy, and capable of the efficient production of both milk and meat.

Since then these cattle, known as Irish Dexters, have been almost crowded out of the livestock picture by larger and more specialized breeds. But here and there some of their descendants continue to thrive.

A Dexter cow at maturity stands only about three feet tall, weighs about 6 or 7 hundred pounds, and requires only half the feed needed by an ordinary dairy cow. But she has dairy capabilities which are impressive for an animal her size.

Robert Weber of Merton, Wisconsin, who has milked a number of Dexters over the years says, "The good ones are really good milkers, but the poor ones are terrible." But his wife takes issue with this. "Oh, that's not true," she says. "He just wants them all to milk like Holsteins. All of ours have milked well." They do agree, however, that the best cow they ever had would give two gallons at each of her two daily milkings.

Dexter milk is naturally homogenized, like goat's milk. Cream will not begin to separate out for a day or two without the use of a separator.

With Dexters, milk production is only half the story. The other half, of course, is meat. Dexters, especially male Dexters, are compact, beefy animals. A mature Dexter bull will generally outweigh the cow by 150 pounds or so, and sometimes may even approach a weight of 1,000 pounds.

Dexter calves weigh about 25 or 30 pounds at birth, but they grow rapidly and fatten well. A Dexter steer will grow, in about fifteen to eighteen months, into a 500-pound animal that will yield a high proportion of quality beef, and does not need much grain to do it.

Being small gives the Dexters another advantage. Their size makes them easy to handle. Dexter raisers tend to make pets of their animals, and it is often the children in the family who care for them and do the milking. They also are capable of fending for themselves and of calving untended without difficulty.

Mrs. Daisy Moore has a herd of thirty-five Dexters on the farm which she and her husband Robert own near Decorah, Iowa. All but half a dozen of these roam for seven or eight months of the year in eighty acres of woods and pasture where they largely take care of themselves. Mrs. Moore provides them with salt, and checks regularly to see that their supply of spring water remains adequate. The rest, including calving, is up to them.

During the four or five months of the year when the pasture is dormant or snow-covered, Mrs. Moore's Dexters are kept in a small field near the barn. They are fed hay—about a third of a bale each per day—but no grain, and despite the severity of northern Iowa winters, she does not find it necessary to provide shelter for them.

A small animal which fares well on a diet composed essentially of grass, produces both rich milk and quality beef, and is hardy and self-reliant yet makes an excellent family pet may seem too much like a homesteader's dream to be quite true. But Dexters are real—and like anything real—owning them poses some problems.

Just finding a Dexter to purchase is likely to be one of them. There are perhaps a hundred Dexter owners in the U.S. Of these, most own only one or two animals. Daisy Moore has one of the largest herds, but she does not often have any animals to sell. She knows of only two or three herds in the country with as many as 20 animals, and puts the total number of Dexters in the country at about 500.

If you would like more information on buying a Dexter, send a stamped, self-addressed envelope to the American Dexter Cattle Association, 707 W. Water St., Decorah, Iowa 52101.

That there are few existing Dexters also means that Dexter breeders work with a limited gene pool, and that makes the matter of maintaining and improving the qualities of the breed more difficult.

When Dexters were first imported, many breeders had difficulty with "bulldogs." These are deformed calves dropped between the fifth and ninth months of pregnancy. This genetic defect is still carried by Dexters. Although it is no longer the serious problem that it once was, the condition still occurs in about 5 to 10 percent of Dexter births. It does not adversely affect the cow, but it is always fatal to the calf. The defect has been attributed to poor interbreeding practices, not the animal.

A Pennsylvania breeder of Dexters, Don Piehota, has calculated from his experience with a herd of twelve animals that a person with three and a half to four acres of pastureland could easily pasture two cows and their calves for about seven months with no supplementary feeding. To keep the animals year-round, Piehota figures expenses of about $180 for 220 bales of hay (one-third bale per animal per day), $40 for one-half ton ear corn, $30 for miscellaneous costs (including vet fees), or a total of about $250 a year. From this investment you would get a heifer ready to be bred at fourteen to fifteen months, worth about $250. The steer at seventeen to eighteen months weighs some 450 pounds and will slaughter out about 250 pounds of meat worth another $250, giving you a return of $500 for your $250 investment, plus the value of the milk.

For those people with limited land and money, and interested in getting enough meat and milk for their family, Dexters may prove to be the ideal animal.

Robert Richardson

Hybrid Buffalo May Be the Best Grazing Animals

For years people have tried to breed buffalo to standard cattle, but the experiments never worked. Whenever the two were successfully mated, sickly, sterile calves were always produced.

The attractive part of such a crossbreed would be the buffalo's sturdiness and ability to put on weight quickly when fed only on grasses, combined with standard cattle's gentleness, and large hindquarters (where the expensive cuts of meat are). After many years of trying, such a cross has been accomplished, resulting in a healthy, true-breeding, fast-gaining hybrid animal now registered as Beefalo.

D. C. Basolo, Jr. spent some 15 years working on a buffalo cross, trying over 1,000 combinations before finding success. He still isn't sure why it works—but it works. The combination he finally hit on is three-eighths buffalo, three-eighths Charolais (a cattle breed developed in France and now popular in the U.S.), and one-fourth Hereford. So fine is the genetic balance that if so much as one-sixteenth more buffalo blood is added, the animal begins to have stockier front shoulders and a thinner rump. Another genetic characteristic of the breed is that half bloods (calves that are half pureblood Beefalo and half other stock) gain weight faster than pureblood Beefalo calves. Most Beefalo in the coming years will be half-bloods, achieved by artificial insemination of Basolo's purebred Beefalo semen with standard cattle, giving an offspring that is three-sixteenth buffalo.

The major claim being made for the new breed is its ability to put on weight without being fed grain. Basolo claims a Beefalo calf will reach slaughter weight of from 1,000 to 1,100 pounds in twelve to fourteen months—*on grass alone*. This is six to eight months faster than grain-fed cattle normally reach slaughter weight. The combination of quicker fattening time and being grass-fed results in cheaper meat, with estimates of up to 40 percent cheaper.

Testing on Beefalo has shown the meat to be about 7 percent fat, with a protein content of 19 to 20 percent, compared to regular meat's 10 to 17 percent protein.

One of the keys to Beefalo's breeding is that it has carried on the buffalo's characteristic of not maturing until about age three. This makes a 1,000-pound Beefalo carcass practically a type of veal. Because the animal is young at slaughter time, the meat is short-grain and very tender, as opposed to a standard breed's mature, long-grain meat. Another benefit of the later maturing age is that Beefalo bulls need not be castrated, because they normally would not become aggressive until age three, well after slaughter age.

Beefalo may offer more bonuses to homesteaders and small farmers than to conventional cattle-raisers. The buffalo used in the original breeding process was a cross between the Plains Buffalo we hear so much about and a lesser-known Woods variety. The resulting animal is about as self-sufficient as an animal can be, regardless of terrain.

Cold weather doesn't bother Beefalo, because their hides have more than 15,000 hairs per square inch, compared to a normal steer's 5,000 hairs to the inch. This gives the animal a fur pelt as opposed to a hide, protecting it in the coldest temperatures. The pelts should have a good market value in the future when more of them are available.

A homesteader or small farmer need not raise

grains for these animals—all they need is pasture—
allowing marginal land to be put to use. In some
tests, Beefalo have performed poorly on standard
forced-grain rations. George Park, sales manager for
Beefalo East, noted that if a Beefalo were given its
choice between corn grain, alfalfa hay, or mixed
pasture grass, the mixed grass would be eaten first,
and corn grain last.

Calves from the hybrids are usually born at from
forty to sixty-five pounds, with a small head and a
long neck, and have shown remarkable ease of calving in all cases.

If you are interested in building up a purebred
herd of Beefalo, it will take about five years of careful
crossbreeding and back-breeding before you have a
true-breeding Beefalo bull. Currently Beefalo semen
is sold by two distributors: Beefalo East, Inc., Route
1, Tazewell, Virginia 24651, and Texas Meat Bro-
kerage, 3800 Beach Road, Burlington, California
94010. Only one purebred bull has been sold to date,
drawing a whopping $2.5 million from a group of
Canadian buyers.

Up to now, all tests involving Beefalo have been
monitored under ideal conditions, and the breed has
done well, if not better than standard breeds. The
animals should come into their own when tested
under adverse conditions, where they are expected
to fare better than standard breeds.

George Park tempered some of the publicity
Beefalo have been receiving by saying, "Under the
worst conditions, you can't expect any animal to hit
1,000 pounds very fast, but these animals will do
much better than what most people have on their
range today. No matter how cold it gets, you don't
have to worry about Beefalo dying. They can live on
almost any type of grass, young trees, or what have
you."

Chapter 9
An Issue of the Day

Organic Farming and National Policy

Not many years ago, an Assistant Secretary of Agriculture scoffed at the idea of an Office of Organic Farming Research within the U.S. Department of Agriculture. Today, he probably is still scoffing, but others are not.

From all sides, the USDA is getting hit for weaknesses in its research program. The National Academy of Science wants a major change in quality and responsibility, while the General Accounting Office criticizes the USDA for ignoring research needs of small farmers. (See following articles.)

Hopefully, the latest NAS report will not be ignored in the same way an earlier critical analysis was three years ago when much of agricultural research was termed "outmoded, pedestrian, and inefficient." Then, as now, the scientists observed in the USDA an unwillingness to plan research needs relating to energy, the environment, and social factors.

An important area of research called for is an "evaluation of alternative technologies for reducing energy requirements in food production and handling." One specific proposal called for a "focus on ways of decreasing dependence upon chemically synthesized nitrogen fertilizer" and to increase reliance on biologically fixed nitrogen by use of manure and inter-cropping with nitrogen-fixing plants.

According to Science magazine, when Sylvan Wittwer, chairman of the Academy's Board on Agriculture and Renewable Resources, was asked if he was advocating a return to the principles of organic farming, Dr. Wittwer replied: "Obviously it relates to the so-called issue of organic farming, but it is broader than that. The use of legumes is becoming a lost technology. That and other techniques of nitrogen fixation are vastly lacking in our nation, and we need to use all the resources we have."

The need to use all our resources obviously starts with the people who now live on farms, and whose numbers continue to shrink rapidly. The GAO study was designed to look at how USDA policies affect small farmers, and it documents just how little help they've been getting.

From 1950 to 1975, the number of U.S. farms decreased about 50 percent. Not only has little research been done for farmers on the low income side, but the overall impact of agricultural research is described as threatening their survival.

Against this background of criticism, there are signs emerging that constructive changes are taking place. Perhaps most simply put, organic matter is taking on new value. This encouraging trend is showing up in a new effort to use organic wastes in agriculture. "Let's not refer to our animal and agricultural residues as waste," declared soil microbiologist T. M. McCalla of the University of Nebraska at a seminar for farmers. "Let's call them resources and start using them."

Identifying himself as an organic farmer, Dr. McCalla told an *Omaha World-Herald* reporter why he advocated the use of natural elements: "We have learned through biological control, that microbes can be used to prevent certain plants and organisms from growing. This means, through biological control, we may become less dependent on chemical fertilizers and pesticides."

The Maine Organic Farmers and Gardeners Association scored a major victory in making organic fertilizer recommendations acceptable, and funding for organic conservation programs available through the Agricultural Stabilization and Conservation Service. After many meetings, phone calls, and letters, a session with University of Maine crop specialists resulted in standards and recommendations for such natural fertilizers as granite dust, greensand, and rock phosphate. ASCS funding is immediately available to Maine organic farmers.

In another development, an agricultural seminar held by the Commission on Maine's Future revealed that Aroostook potato farmers were "wearing out their land by over-use and hard chemicals." A summary of the session in *Maine Times* reminds one of a typical meeting of an organic farming group, during which comments like "producing more of our own feed grains," "the state must promote crop rotation and seek new fertilizing methods," and "Maine is ideally suited for smaller, low-energy organic farms" were made. But now these observations are expressed at official state-wide policy meetings.

With increasing frequency, scientists are documenting the need for the kinds of changes sought by organic growers. In a follow-up report to his original research, Dr. David Pimentel and associates at Cornell have published "Energy and Land Constraints in Food Protein Production." After many references to greater human consumption of vegetable protein, energy loss in feedlot cattle production, the need to preserve farmland, and people to work the farms, the authors note that crops have physiological limits in their ability to respond to increased amounts of fertilizers. There are limits to technology just like there are limits to growth, and Pimentel quotes Sir Julian Huxley who argued that science has been "completely unable to cope with the appalling problems" of the developing world today.

The people in charge of agricultural research in the U.S. have a tough song-and-dance act to perform. One tune goes like this: "Since 1950, the farmer's output per manhour has gone up 6 percent per year and he now feeds himself and fifty-six others." But the dance step and lyrics have a jarring message about "half the farmers are gone, and energy reserves are dwindling, prices are high, and cancer rates from environmental factors are soaring."

For a great many experts, organic farming methods will put us back fifty years. "We cannot go back" says Crops & Soils Magazine published by the American Society of Agronomy. They ask (and answer): "Where are the 61 million horses and mules that would be needed to sustain present agricultural production? Only 3 million exist in the U.S. today."

But horses and mules, while compatible with organic methods, are not the essential ingredients of an organic agricultural system. Tough as it may be for the American Society of Agronomy and Earl Butz to comprehend, many organic farmers use tractors and modern tillage equipment. They are proud of the profits they make from their integrated operations, are proud of the fertility of their soil, and are proud that their children will eventually take over their farms.

The major difference between organic farmers and traditional agronomists is not over going *back*. Rather the difference lies in how we go *forward!*

The indictments of the USDA handed down by the National Academy of Science and the General Accounting Office focus on glaring weaknesses of present U.S. farm programs. Another indictment of how these same approaches are wrecking farmers in other nations appears in the British publication, *New Scientist*, under the whimsical heading, "Agronomy to the Rescue."

As seen by Dr. Charles Posner, Guatemala has raised its per capita food production by 11 percent since 1960, but the sad fact is that mass starvation is now a possibility there. The paradox is primarily attributed to a handful of native agronomists who were schooled in U.S., European, and Soviet practices. They all argue that large-scale farming is more efficient than the family farm and that specialization means a higher return than integrated farming. "Following this theory," writes Dr. Posner, "the agronomists encouraged turning over the richest agricultural land in the country for grazing, and also encouraged the clearing of jungle for plantations which turned the area into a virtual desert."

Now a study in Guatemala shows that the small farm is more efficient than the large farm. "Farms of 1 acre have maize and wheat yields no lower than farms of more than 1,500 acres, yet the energy input on the larger farms is three times as high as the tiny farms," explains Dr. Posner. "Yields for rice, beans, potatoes, and other staples are between 10 and 60 percent higher on the small farms."

From advocates of organic methods the world over, the cry is not to ignore expert agronomists but instead to reeducate the agronomists and come up with a "new" agronomy.

The "new" agronomy is actually well-launched, albeit disconnected. At just about every agricultural experiment station, projects which are vital to organic agriculture are underway. But only in rare instances are "high science" and "people benefits" matched up.

Innovative methods are also being sought to save our soils. One plan, originated by Wesley Buchele, a professor of agricultural engineering at Iowa State University, calls for a *compulsory* system of soil conservation. ("Land is for living, not the basis for accumulating wealth.")

Prof. Buchele maintains that our nation must create a "conservation group" since forty years' results of Soil Conservation Service efforts have demonstrated that voluntary soil conservation programs have not prevented the deterioration of the land. In 1935 three billion tons of soil were lost from cultivated lands in the U.S. After spending more than $11 billion in thirty years, four billion tons of soil were lost in 1964. According to Prof. Buchele, more than ten pounds of soil are lost for each pound of grain harvested.

For a modest number of people who make their living in farming, organic agriculture has provided the *voluntary* incentives to use methods that save soil and yield satisfactory harvests. It seems more than time for our nation's policy-makers to get off of their pedestals and begin to analyze what organic farmers have been doing right, and why they have been doing it!

The fact is that organic farming has a great deal to offer for the better future of all of us.

Jerome Goldstein

Small farmers currently exist on a profitable basis, despite the lack of research directed towards their needs. The GAO has highlighted this need for research and the USDA's failure to provide these farmers with information.

Organic Farming

GAO Documents Need for Small-Farm Research

Farming is a highly competitive and risky industry. Each year fewer people earn their livelihood by farming. The trend has been toward fewer farms that are larger and highly mechanized. In today's market farmers who do not increase the size of their farms, mechanize their operations, or otherwise update their farming practices soon become noncompetitive and eventually drop out of the mainstream of farming.

To help farmers produce and market food and fiber efficiently, the U.S. Department of Agriculture (USDA) has, over many years, carried out and helped finance agricultural research and extension activities aimed at gaining and applying knowledge and technology efficiently to the biological, physical, and economic phases of producing, processing, and distributing farm and forest products.

In recent years, much concern has been expressed about whether enough of this research and extension activity has been directed to the problems and operations of the small-farm operator. A report by the General Accounting Office (GAO) to the congress in 1975 discussed some of the problems impeding the economic improvement of the small-farm operations and research and extension efforts of USDA and land-grant colleges for improving the farming operations of small-farm operators. The below information is from the GAO report.

Although some publicly supported extension and research projects have related to the needs of small-farm operators, USDA and land-grant colleges have not made a concerted effort to solve problems impeding the economic improvement of small-farm

308

operations. Also, they have not adequately evaluated the economic and social impacts of production-efficiency research nor determined the assistance that small-farm operators need to plan for and adjust to changes brought about by such research.

Generally, farms with the least amount of farm sales have gone out of business. As shown in Table 9-1, for example, the number of farms with gross annual sales under $20,000 decreased by about 1.8 million between 1960 and 1973. Because of increased prices, production efficiencies, and farm size, about one-third of these moved into the category of farms with gross annual sales of $20,000 and over. The other two-thirds, however, went out of business.

TABLE 9-1

Gross annual sales	Number of farms 1960	1973	Percent of change 1960-73
Expanding farm sector:			
$40,000 and over	113,000	446,000	294.7
$20,000 to $39,999	227,000	563,000	148.0
Total	340,000	1,009,000	
Declining farm sector:			
$10,000 to $19,999	497,000	332,000	-33.2
$ 5,000 to $ 9,999	660,000	262,000	-60.3
$ 2,500 to $ 4,999	617,000	488,000	-20.9
Less than $ 2,500	1,849,000	753,000	-59.3
Total	3,623,000	1,835,000	

Although the number of smaller farms has declined and many of those remaining are operated by part-time or semiretired farmers, many small farms are operated by farm families in their productive years who depend primarily on farm income for their livelihood.

For this review, we considered a small-farm operator as a person who 1) is under sixty-five years of age; 2) works off the farm for wages less than 100 days a year; and 3) sells less than $20,000 of agricultural products annually.

Using information from the 1969 Census of Agriculture on the profile of farmers, we estimated that 666,000, or about 37 percent of the operators of the 1.8 million farms with agricultural sales of less than $20,000 in 1973, met the above definition of a small-farm operator.

The basic responsibilities for agricultural research are set forth in the Organic Act of 1862 which established USDA, and the Hatch Act of 1887, which established state agricultural experiment stations at land-grant colleges.

USDA and the land-grant colleges have defined agricultural research as a systematic method of gaining and applying knowledge efficiently. The research is divided generally into four categories: production efficiency, marketing, foreign-oriented, and people-oriented. The primary link between agricultural research and the farmer is the cooperative agricultural extension agent. The basic mission of agricultural extension is to help people identify and solve their farm, home, and community problems through use of research findings and USDA programs. Land-grant colleges carry out extension work through state and county extension offices. The states employ county, home economics, and 4-H club agents, state and area specialists, nutritional aides, and others to conduct educational programs adapted to local problems and conditions.

Although some publicly supported extension and research projects have related to the needs of small-farm operators, USDA and the land-grant colleges have not made a concerted effort to solve problems impeding the economic improvement of small-farm operations. USDA and the land-grant colleges have not, to a great extent, evaluated the economic and social impacts of production-efficiency research nor determined the assistance that small-farm operators need to plan for and adjust to the changes brought about by such research.

A large amount of the nation's agricultural land is under the management of small-farm operators. Much of this land is not being used to its full production potential. Demonstration projects conducted by the cooperative extension organizations have shown that some small-farm operators are capable of increasing the productivity of their land and increasing their incomes if they are helped to do so through more intensified extension and research efforts.

The value of using available resources efficiently was demonstrated in a University of Minnesota study comparing the earnings in 1971 and 1972 of dairy farmers with herds in given size categories: twenty-five to thirty-four cows, thirty-five to forty-four cows, and forty-five to sixty-four cows. The study showed that, in terms of labor earnings, the top 25 percent of the farmers in these categories earned from 4.9 to 7.8 times more, on the average, than the bottom 25 percent, although both groups had similar kinds and amounts of resources.

The study's authors concluded that the efficient use of similar amounts and kinds of resources was the major reason for the large variations in the farmers' earnings.

Extension Service officials said that their experience showed that small-farm operators can improve their operations by effectively using available technology and efficient management practices. Cooperative extension organizations have, for a number of years, sponsored demonstration programs, some jointly with the Tennessee Valley Authority, to extend training and technical assistance to small-farm operators.

The type and intensity of assistance provided and resulting accomplishments widely differed between programs as did the abilities and resources of participating farmers. Nevertheless, the results of these programs showed that there are small-farm operators who will respond to extension efforts specifically designed to meet their needs and who can be helped to increase production by using available technology and efficient management practices. Three of these programs are discussed below.

THE RAPID-ADJUSTMENT FARM PROGRAM

This program is designed to assist the farmer who has the potential to become a full-time commercial farmer and earn a satisfactory income. The Tennessee Valley Authority and various extension organizations in the Tennessee Valley cosponsor the program.

After a farm is selected for participation, all phases of the farming operation are surveyed. Benchmark data is collected on land and land capabilities, labor supply, management ability, livestock, machinery, equipment, supplies, financial statements, and credit availability. The operator's management ability is evaluated mainly on past and present use of resources, crop yields, livestock production levels, credit uses, and community leadership. On the basis of this data, alternative farm plans are developed for the farmer and presented to him. He selects the plan to implement.

After a final plan is agreed upon, the local extension agent and land-grant college specialists work closely with the farmer to help him overcome problems impeding implementation of the plan.

A report by a Tennessee Valley Authority official and a University of Tennessee professor in February 1972 showed that, for the sixty-two farmers in seven

312

Tennessee Valley states who had completed four years in the program at the end of 1970:

Average farm investment had risen from $48,707 to $70,080, an increase of 44 percent.

Average gross sales had risen from $18,026 to $31,493, an increase of 75 percent.

Average net farm income had risen from $4,548 to $7,935, an increase of 74 percent.

THE ELK RIVER PROJECT

In 1959 the average size of the 12,260 farms in the Elk River area (a seven-county area in south central Tennessee) was 112 acres with average farm sales of $2,862. The University of Tennessee, the Tennessee Valley Authority, and local groups studied the problems impeding the development of agriculture in that area and evaluated the improvement potential.

Programs were established during the 1960s to accelerate the use of recommended crop production practices, improve livestock production practices, and improve farm management practices. The primary goal was to increase gross farm sales to $50 million by 1970—an average annual growth of $1.6 million or over twice the thirteen-year growth trend of $700,000.

Examples of reported accomplishments were:

Sales from livestock and livestock products increased about 75 percent.

As livestock production increased, better marketing facilities were established and farmers received better prices. In 1969 feeder pigs sold for about twenty-one dollars a head. Before organized marketing facilities existed in the area, feeder pigs sold for ten dollars a head.

Net annual income per farm increased from a 1959-61 average of $1,773 to a 1967-69 average of $2,860—a 61 percent increase.

THE TEXAS INTENSIFIED FARM-PLANNING PROGRAM

This program, which used local farmers as program aides, was designed by the Texas Agricultural Extension Service to show the effectiveness of working with small-farm operators on an intensive basis to change production and management practices, and to provide county extension staffs an opportunity to field-test program procedures, teaching methods, and techniques which could be drawn upon to strengthen an educational program designed to assist small-farm operators.

On the average, the 224 participants were fifty-four years of age, operated 121-acre farms of which 100 acres were used for pasture and 19 acres for cultivation, and earned $1,828 from the sale of farm products in 1968, the year preceding the start of the program. They produced and sold beef cattle, swine, corn, cotton, grain sorghum, peanuts, watermelons, peas, cucumbers, potatoes, tomatoes, and cantaloupes.

A report on an evaluation of the program by Texas A&M University showed that, on the average, the gross income from livestock sales increased from $1,111 in 1968 to $1,389 in 1970—an increase of about 25 percent—and the gross income from crop sales increased from $1,082 to $1,089—an increase of less than 1 percent. Although it did not estimate the value of each factor, the report indicated that improved pastures, livestock production practices, herd expansion, and calf-crop percentages and higher prices could account for the increases in livestock income. The evaluation team concluded that program aides greatly helped participants increase their livestock income.

In a September 1969 speech before the Division of Agricultural and Food Chemistry of the American Chemical Society, Ned Bayley, the former Director of USDAs Office of Science and Education (now the Office of the Assistant Secretary for Conservation, Research, and Education)—USDAs focal point for coordinating research policy development, planning, and evaluation—said:

> When we ask what agricultural research has done for this group of farmers (small-farm operators), the answer comes back: "Very little." In fact, the overall impact of agricultural research has threatened their survival.

Bayley also said that outstanding social and natural scientists, who were sensitive to the problems of small-farm operators, should be brought together to thoroughly delineate researchable areas where answers could be found to their problems.

The former director told us that the study group he had suggested had never been brought together and that the points he had made in the 1969 speech were still valid.

To more fully achieve the potential national and individual benefits of extension and research programs aimed at encouraging and helping small-farm operators to improve their farming operations, we recommend to the Secretary of Agriculture that USDA:

> Identify small-farm operators in their productive years who depend on the farm as their primary source of income and categorize them according to their resources, abilities, educational experiences, and willingness to improve their operations by using available

technology and efficient management practices.

Estimate the costs and benefits of programs needed to extend training and technical assistance to small-farm operators having the potential for improvement and present the information to the Congress for its consideration.

Examine the potential for research uniquely designed to improve the economic position of small-farm operators and, if such potential exists, consider the priority of such research in relation to other federally funded agricultural research.

Establish prodecures for: 1) evaluating the economic and social impacts of future research that could greatly change the productivity, structure, and/or size of existing farms; and 2) determining the assistance small-farm operators would need to plan for and adjust to the resulting changes.

What Kind of Money Can a Small Farm Make?

In West Virginia, folks take their small farmers (anyone selling $500 or more of produce annually) seriously. "Of course we think small farmers are important," says Gus Douglass, State Commissioner of Agriculture. *"That's what 80 percent of our food producers are."*

West Virginia, as far back as 1950, started a policy to look out for small growers. The first effort was

"Farmers Market Programs" to provide a means by which small farmers could consolidate and market their produce. Initially a wholesaling effort, the programs now promote retailing, too, not only at the farmers' markets, but they also provide individual farms with up-to-date information on running roadside stands and pick-your-own operations.

Not content with just "paper help," state officials decided to put their belief in small farming on the line. Near Hamlin, an eight-acre site was developed into a small demonstration farm. Objective: Show small growers how to make money with small-scale market gardening. Operated cooperatively by the State Department of Agriculture and the University of West Virginia, the project nevertheless emphasizes practical, real-life methods a small grower can financially afford. There's nothing fancy about the farm; no big money flushed into it, nor free student labor as you might find at most university farms. Equipment is what the serious small grower needs and no extras. Every hour of labor is religiously recorded. In other words, a small grower can visit the farm and go away convinced he can do what's being done there. Or better. "One fellow studied our greenhouse bedding plant production, went back home and started his own. He was netting $600 a week last time I heard from him," says Ralph Swanson who manages the farm. "That's better than we've been able to do so far."

The little grower must sell direct to consumers if he is to profit from his comparatively small amount of produce. So the demonstration farm sells retail, too. Some of the crops are harvested and then sold; some by the pick-your-own method. "We had anticipated more marketing problems than we experienced," says Kemper. "There's been no trouble at all selling all the corn, watermelons, strawberries, tomatoes, bedding plants, and flowers we produced. Even in this fairly out-of-the-way location, we've never had to advertise. Our only failure was with pick-your-own snapbeans. People weren't interested so we quit growing them."

The real value of the demonstration farm for mini-farmers is the way it uses labor as its basic economic yardstick. The researchers realized that labor for the very small grower is not the same as labor for the larger grower. For the latter, labor is all cost, a significant cost if he is very large, a minor cost if he can replace most of it with expensive machinery. For the small grower, labor is a much more personal thing; usually it's all his and his family's. It doesn't cost him cash out of his pocket. He may even think of labor partially as hobby or recreation, but in any event, it is by far the largest input into his little operation. *For him, knowing what crops give him the best return per hour spent can become the single most important step in planning his mini-farm.*

The Hamlin Demonstration Project doesn't have all the answers yet, but its operators have learned a thing or two. Already disproved is the traditional mid-South belief that a small tobacco allotment is one of the better cash crops for small farms. "Sure, you can sell $600 worth of tobacco off a quarter-acre plot," says Swanson, "but it takes 134 hours of labor. After you pay your other costs, you'll find your return to labor on a small tobacco plot is around $3 per hour. A quarter-acre planting of strawberries sold pick-your-own returns over $10 per hour."

But while strawberries have so far proven to be the best crop on the farm from the standpoint of return to labor, they were not the most lucrative. "Tomatoes are our main crop, and we plan all our other production around them," says Swanson. "The return per man-hour is only around five dollars to six dollars after expenses, but the marketing season is much longer."

The farm used 200 hours of labor on one and one-fourth acres of strawberries to gross $2,600. But with 400 hours of labor on tomatoes stretched over three months instead of the one for strawberries, the farm grossed nearly that much money on only one-half acre.

Watermelons and sweet corn are the other two crops the farm has found to have sufficient appeal to consumers for successful direct selling. "From a quarter-acre plot of corn or melons, you can net $100 which doesn't seem much after the berry and tomato sales. But the melons and corn require only eight hours of labor each, or a return per man-hour nearly as high as strawberries," says Swanson.

The researchers believe apples and grapes can be profitable for small growers. Both require several years before they produce income, however, making them a little more of a gamble than vegetables. The farm's acre of fruit trees was planted two years ago; the grapes went in last spring. The apple trees—all RED DELICIOUS and GOLDEN DELICIOUS because of their proven reliability and marketability—are dwarfs trained to a four-wire trellis. "This is a typical hedgerow planting, 500 trees to the acre. The trees will be maintained at about seven feet in height for easy, fast picking," says Kemper.

Below is a summary of labor requirements and return per man-hour on various garden crops based on the Hamlin data. We have reduced plantings to quarter-acre plots (a plot about 108 by 100 feet) for convenience; if your plots are smaller or larger you can do your own division and multiplication. The costs are estimates by Hamlin researchers and do not include a charge for the land or depreciation of machinery, both of which would not be particularly important in most mini-farm operations anyway. Costs include a moderate amount of chemicals since the demonstration farm is not totally organic. Organic growers will have lower costs in this regard but usually higher labor in applying organic fertilizers, so that the end result in return per man-hour should be about the same. Where the organic grower can gain, however, is in long-term buildup of organic matter and fertility. The chemical grower will have to continue to apply chemicals, if not at a higher rate, then at a higher price, while the organic grower may reach the point where his soil needs no additional fertility.

Add six dollars per quarter-acre cost if you use drip irrigation. "We believe that irrigation is important for the serious market gardener, even the smallest," says Swanson. "Sprinklers are pretty expensive, but plastic soaker hose can be installed for about twenty-five dollars an acre if you've got a water spigot handy." The soaker hose is Chafin Twin-Wall hose, really a hose inside a hose. It truly is an intriguing arrangement that guarantees equal distribution of the water along the entire length. The water drips very slowly through minute holes in the outer hose. Ideally, the hose is positioned along the base of the plants and covered with mulch. A little water goes a long way with this arrangement.

¼-acre	Hours of labor	Gross sales	Estimated costs	Approximate return per man-hour
Tomatoes	200	$1,200	$200	$ 5
Strawberries	40	$ 530	$ 40	$12
Sweet corn	8	$ 100	$ 12	$11
Watermelons	8	$ 100	$ 20	$10

Consider the accompanying figures as guidelines only. You may do better or worse. Variations in prices, soils, consumer preference, etc. can alter the figures up or down.

Tomatoes were trellised, pruned, and mulched with plastic. Watermelons were mulched with plastic, strawberries with straw. All crops were sold at current fresh market prices for 1975.

Some deductions: Peppers and eggplant, while grown like tomatoes, require no pruning or staking and not as much harvest time. Figure eighty hours of labor per quarter-acre. Bramble berries actually work out to about the same figures as strawberries. You don't have to plant a new crop every year (raspberries should bear well for ten years) but you have more pruning and trellising. Bramble berries don't produce the poundage per acre that strawberries do, but sell for a higher price. Muskmelons, squash, cucumbers, and other vining plants that grow like watermelons will take the same number of hours of labor as the latter. Muskmelons should gross as much or more, especially in the North. But cukes and squash will obviously sell for less—if at all.

Can an eight-acre mini-farm with a greenhouse make a good family living? The researchers think so. Already the little farm has grossed $12,000, and with more know-how, and full production from the apples, grapes, and greenhouse, they think that gross can be doubled, at least. But even as a part-time activity, small-scale market gardening which substitutes owner labor for expensive machinery and chemicals can return ten dollars per hour.

Gene Logsdon

An Organic Training Farm

A 650-acre farm in rural Anson County, North Carolina, is the home of the Frank P. Graham Experimental Farm and Training Center. It's unique among the 2,800,000 farms remaining in the United States (the lowest number since record-keeping began in 1910). While the farm produces a delicious variety of vegetable crops, its primary product is people.

Five hundred and eighty acres of abandoned farmland were bought in 1972 by the Rural Advancement Fund of the National Sharecroppers Fund, Inc. An additional 144 acres of land were bought in 1975. A complex of fields, work buildings, and dormitories was quickly developed, and today it stands as the only facility in the country, public or private, offering a free training program to disadvantaged rural people who want to make a go of it on the land, or to urban people who want to return to the land.

Poor black and white farmers, Chicano migrants, and day laborers learn how to farm without using expensive chemical fertilizers, insecticides, and herbicides. This emphasis on natural farming isn't a faddish cashing-in on the current popularity of organic foods, but a hard-headed, conservative commitment to time-tested techniques that offer an alternative to rural despair. The Center's beans, squash, tomatoes, yams, peas, watermelons, lettuce, and cucumbers stock the farm's own larder, and bring profits from demanding markets.

While the trainees have been learning by producing in the fields, they are being taught other skills such as carpentry, masonry, welding, bookkeeping, and equipment repair so they can supplement farm incomes by mastering trades needed in their communities. They are also taught how to organize and manage rural cooperatives.

National Sharecroppers Fund Executive Director Jim Pierce, who pioneered the development of the Center, says, "For too many years USDA's conventional wisdom has been that poor, small farmers are expendable, that nothing can really be done to make marginal operations pay, and that the future belongs to agribusiness.

"Farm numbers have been steadily declining since 1936. We lost another 14,000 last year. Agriculture Secretary Earl Butz says that by 1980 we will have a *million fewer farmers* in this country."

An Oklahoman who is part Cherokee Indian, Pierce grew up on a hardscrabble farm during the Depression. Then, for a quarter of a century, he helped to organize workers in southern industry. He saw the pattern repeated time and time again, of people squeezed off farms into menial factory and mill jobs, of land laid waste by poor farming practices.

"We offer a total approach to rural problems here: how to farm most economically and productively on a small holding; how low-income farmers can link up with others like themselves to form cooperatives which are the best single hope they have. We teach how co-ops can hold their own in the tough, tricky world of marketing," Pierce says.

When the RAF bought the Anson County property, the land was overgrown and the soil exhausted from decades of cotton sharecropping. It's the only kind of land that many poor farmers ever get their hands on.

Charles Dixon, the Center's farm manager who had previously organized a farmers' cooperative in Virginia, needed help clearing and leveling the wasted acres. One day, a local man by the name of Bennie Gaddy came by to see what all the activity was about in the fields surrounding the shack. "I'd quit farming because I couldn't support my family and was working in a textile mill. But I didn't want that. I wanted to farm. So I stayed on at the Center."

Dixon and Gaddy felled trees, pulled stumps, cleared brush, and started what would become massive compost heaps. They plowed into the ground a "home-brewed" fertilizer mix of feather meal, seaweed, ground granite, bone meal, tankage, and other soil-conditioners. By the spring of 1973, land that had been written off as finished had a new lease on life—fields of beans, potatoes, lettuce, and beets moved visiting local farmers to cautious words of country praise.

Before trainees could come in, the physical plant had to be completed. Within a year, all the basic structures were built. The main building exemplifies the character of the Center. Sprawling like a brown brick ranch house, it is attractive, homey, and practical. Housed in it are Center offices, dormitories, kitchen, a fifty-seat dining room, and recreation room. The dorm, which can accommodate thirty-two persons, is divided into rooms with bunkbed space for six or eight, with individual bureaus and curtained windows.

One of the Center instructors, who came off a small farm to earn a degree in agriculture, explains the Center's recipe for better farming: "What we put into the fertilizer is determined by soil analysis. Depending on the soil, we'll use more of one thing, less of something else. But no chemicals. And we've yet to use any pesticides on a crop here. We plant a winter-cover crop of rye, clover, and cowpeas, which puts nitrogen back into the soil naturally. Last year, trainees even milled seed from their own rye grass."

The instructor is a tall, serious young man not given to enthusiasm. Asked about the commitment of the trainees he teaches and the future of the Center, he replies: "Everyone out here is looking for something. Some are idealists who like the notion of natural farming but don't like to get their hands dirty. So the first thing we do is burst some bubbles and set realistic goals, because vegetable farming involves a lot of hard, consistent labor. I'm not much of an idealist myself, but I hope the Center prospers. They're looking for answers to problems that most people only talk about. Well, they're well past sitting and talking about it. They're getting their hands dirty and working up a sweat, and as much will come out of it as the people will put into it."

The Graham Center is financed by individual contributions to the Rural Advancement Fund which are tax-deductible. The address of RAF is Room 100, 2128 Commonwealth Ave., Charlotte, NC 28205.

Paul Good

Helping Missouri's Small Farmers

When Ron Sayre moved his wife and two children to a hilly eighty-acre farm near Unionville in northwest Missouri four years ago, he was typical of the kind of young farmer the experts say will never make it.

The money he saved during seven years in the Navy had been used in buying the farm and a few milk cows. And he farmed only part-time while commuting to Northeast Missouri State thirty-five miles away for classes in agriculture.

At the other end of Putnam County, Sam Valentine and his family were working a small livestock farm he had bought several years earlier. Valentine, an evangelist preacher, had returned to the area at the age of forty-two after several years in Chattanooga, Tennessee.

He was considered a poor prospect, too, in a production system that had become increasingly competitive. It was assumed by most agricultural policymakers that small farmers like Sayre and Valentine were on the way out and that "efficient" operators with big farms and big machinery would soon produce all the nation's food.

There was considerable debate over this assumption at the University of Missouri, however, with several extension specialists insisting small farmers should not be written off. They had data, in fact, that showed many small operators were continuing to move *into* agriculture.

This debate was more than academic because nearly 100,000 of Missouri's 137,000 farms at that time were reporting less than $10,000 in annual sales. Records showed production expenses took 70-80 percent of this gross income, leaving these small farmers with no more than $3,000 a year for living expenses.

The result was that Missouri late in 1971 launched its Small-Farm Program in Wayne and Polk Counties. When it was clear it would be accepted and supported locally, it was expanded to include Putnam and five other counties scattered across the state.

"After completing our initial survey, we made personal calls on 451 families in Putnam County that were in the small-farm category," Burch Harrington, farm management specialist in the Green Hills Extension Area, recalled.

"A total of 110 families wanted to participate so we hired Lewis Harbert and Jay Ross, both small farmers in the county, as education specialists to work with them."

Sayre and Valentine were in the initial group that Harbert and Ross worked with, discussing problems during regular farm visits and providing information needed for decisions on everything from selecting garden seed varieties to building portable farrowing crates.

"Lack of capital was their main problem and most of them had neither money nor the knowledge about where to go to get it," Harrington explained. "Some also were reluctant to operate on borrowed money so we helped them do a better job with the resources they had."

Nearly 900 families living on small Missouri farms have been in the program since it began. Although their farms averaged only 153 acres, and much of it was pasture and hay ground, nearly all owned most of their own land. Their average age was forty-seven, at least five years under the average for all farmers in these counties and much younger than the usual stereotype of small or part-time operators.

The University of Missouri's latest report, which covers the program's first three years, shows most of these small farmers increased net income by expanding livestock operations. This provided better utilization of both family labor and of pasture, hay, and other available resources.

The most impressive livestock expansion was on thirty-eight small dairy farms, where the cows-per-herd average went from thirteen to twenty-nine. Records show 57 percent of these operators also increased production-per-cow.

Sayre beat that average during his three years in the program, increasing the size of his Holstein herd to thirty-two, planning and building a modern barn and milking parlor himself, and adding 131 rented acres to his basic 80-acre unit. He also completed work on his B.S. in animal science so he could farm fulltime.

His herd production average is more than 12,000 pounds a year now and his carefully kept records show one of his cows produced 19,000 pounds of milk the first 319 days of 1975. "All my cows are producing 12,000 pounds and up except four," he said, "and I'm culling them out of the herd."

By renting land adjoining his own, Sayre produces enough corn for silage for his growing dairy herd, and some soybeans for a cash crop. He still has less than $2,000 invested in farm machinery and gained so well financially that he recently obtained a loan for a new house, keeping the mortgage low by doing the painting and other finishing work himself.

The Missouri report showed small farmers with hogs averaged nine sows per farm the year before participating, compared with a thirteen-sow average three years later. Feeder pigs sold per farm increased from 123 to 141.

"The average number of beef cows increased from fifteen to twenty-four per farm at the end of 1974, with most of this increase coming from calves they had raised," the report added. "Most farms provided additional pasture and hay needed for the expanded cow herd by improving and using grass on land under their control."

Valentine has increased his net income by expanding both his cattle and hog operations. He farrows twice a year, now averages eight pigs per litter, and gets a 95 percent calf crop.

Harbert, who has worked with Valentine the last three years, is proud of the way he has experimented with double-cropping to increase his grain output on limited acreage. He had one field last year that produced forty-four bushels of wheat and 111 bales of wheat straw per acre, followed in the fall with a crop of soybeans that yielded twenty-two bushels per acre.

"He's getting marvelous production off this farm," Harbert said.

Are Sayre and Valentine exceptions in the group of small-farmer participants and are they doing better than other farmers with the same size operations and limited resources?

The University of Missouri, in a recent evaluation study designed to answer questions like these, came up with this significant conclusion:

"Participants in the small-farm program had

higher farm sales, higher net farm income, larger enterprises, more livestock assets, slightly more efficient resource utilization, more improvements in housing, and more stability in level of production."

One important difference was that participants kept dairy and beef cows when prices declined in late 1974 and throughout 1975, while there was considerable reduction in herd size on nonparticipating farms.

"Sow numbers were reduced on both but there was a tendency for participants to keep gilts, which was not apparent on the nonparticipating farms," the evaluators found. "This placed participating farms in much better position to benefit from increasing prices for feeder pigs in 1975."

Thus the program apparently reduces the tendency to expand when prices are high and cut back production when they decline. This "inner-outer" problem contributes to wide livestock price swings and often leaves small producers with too many hogs or cattle to sell when prices are declining and few or none to sell when they start going up.

It isn't known when, or if, funds will be available to expand Missouri's small-farm effort more than adding a new county here and there. Mercer, the county west of Putnam, is starting the program this year under the experienced direction of Harrington and his assistants.

There is no lack of enthusiasm for it among the top brass at the University of Missouri. Schell H. Bodenhamer, associate dean of the College of Agriculture, is sold on both expansion and the need for separate extension programs for commercial farms and for small farms.

"Allocation of limited extension resources between commercial and small farms . . . is a continuing decision," he said in a recent statement.

"Funds for an additional 100 small-farm educational assistants (enough for 5,000 small-farm participants) would be an excellent investment in Missouri's future."

Roger Blobaum

Future Developments or Farms?

If you drive across the Midwest on Interstate 80 in early August, when rows of deep green corn and soybeans stretch as far as you can see in all directions, you can't help but feel secure about the future of America's food supply.

This agricultural abundance is especially reassuring at a time when global population growth is outstripping food production gains, the number of grain-exporting nations has dwindled to seven, and world population estimates of 6.5 to 7 billion by the year 2000 are widely accepted.

Although farmers have been increasing food production regularly since the 1940s, and routinely produce much more than the nation can consume, there are indications they are approaching some production limits. American agriculture, while appearing to be invincible, needs help from farm and city people alike in solving a wide range of problems.

Continued conversion of millions of acres of cropland to non-agricultural uses, even in farm states like Illinois and California, is one of the most pressing problems. A recent Senate study concluded that urban development will consume additional land equal to the total areas of New Hampshire, Vermont, Massachusetts, and Rhode Island by the year 2000.

Another growing problem is topsoil erosion, soil salinity, and other resource deterioration that indicates agricultural land is being abused. A third of the nation's topsoil is gone already, according to an Academy of Science report, and a recent inventory shows only half is adequately protected against erosion.

New on the problem list are energy and fertilizer shortages, depletion of ground water used for irrigation, tremendous expansion in energy development, and reports indicating agricultural growth is leveling off after thirty years of uninterrupted productivity gains.

These trends are being monitored by such academic leaders as Dean Anson Bertrand of Texas Tech University's College of Agriculture, who warns that current world population growth requires a food production increase more than double that of recent years.

"The situation is critical now," he said in a recent speech, "and may reach catastrophic proportions by the year 2000."

Although new cropland is regularly added in some areas, mainly through irrigation and drainage, urban growth and other development consume much more. The net loss is about 1.4 million acres a year and rising, with farms near cities and cropland classified as "prime" disappearing fastest.

Only about 47 million acres of Class I land, the kind usually referred to as "prime," is left. These soils are deep, well drained, resistant to erosion, and easily worked. They also are highly productive, suitable for intensive cropping, and irreplaceable.

It is estimated that about 17.2 percent of all farms are in the nation's 242 Standard Metropolitan Statistical Areas, usually referred to as SMSAs, which puts them directly in the path of urban expansion. These farms produce about 21 percent of the value of all agricultural products sold and about a fifth of the nation's food. The Urban Land Institute estimates that the land area of these urban regions will reach 486,902 square miles by the year 2000, nearly a sixth of the nation's total land area.

This pattern of cropland conversion, plus air pollution and other urban-related problems, makes it increasingly difficult to produce fresh vegetables and fruits near metropolitan areas, where truck farms and orchards traditionally have operated. The Economic Research Service reports 60 percent of all vegetables still came from urban areas as recently as 1969.

Not much is being done about this. The Citizens Advisory Committee on Environmental Quality, in a recent report to President Ford, pointed out that the United States has no policy or plans for preserving agricultural land and leaves most of the decisions to speculators, developers, and others who view land as a commodity.

"Our study indicates that more than 54 million acres of cropland were converted to irreversible uses in a recent twenty-year period," the committee states. "Most of this land was used for urban housing, highways, airports, power plants, waste disposal sites, shopping centers, and reservoir construction."

The most urgent need would seem to be preserving land in regions like the Corn Belt where ample rainfall, rich soil, and ideal growing conditions virtually assure good crops. Farmers can produce more food for less there than where the growing season is short, soils are difficult to manage, or irrigation is needed.

Preservation of good land close to cities also is important to maintain future options for land application of city wastes, more locally grown produce, and other food system changes that make sense in an

332

energy-short economy. These include development of direct farmer-consumer ties through farmers markets and food cooperatives.

The potential for applying wastes to cropland near cities becomes increasingly attractive as fertilizer shortages develop and prices go up. Research in several areas shows applications of sludge and other wastes can supplement fertilizer applications, improve soil structure, and help crops survive drought.

Yet real estate developers prefer the same level, well-drained tracts on the urban fringe that offer economically sound possibilities for city waste application. They meet little resistance in most areas in turning orchards, truck farms, and other agricultural units into shopping centers, subdivisions, and other permanent development.

Developers also have invaded areas producing artichokes, avocados, and other specialty crops that have unique soil and climate requirements. They are often grown where weather is tempered by large bodies of water, the same areas sought for subdivisions, recreation complexes, and second-home sites.

Probably the most serious consequence of conversion of high-quality cropland, is the pressure it puts on food producers to plow up grazing and hay ground that is much more difficult to manage and a lot less productive. The nation has more than 100 million acres of marginal land, mostly in trees and grass, that is not considered suitable for regular cropping.

Few people are more aware of the implications of marginal land conversion than Administrator Mel Davis of the Soil Conservation Service. He is urging soil scientists, both in SCS and elsewhere, to discourage farmers from bringing into production land that cannot be protected against erosion at reasonable cost.

"If 100 million acres of Classes I-III land were converted to cropland in the next ten to fifteen years, additional soil loss could be as much as 670 million tons per year based on experience in 1973-75," Davis reported. "Such accelerated pollution and destruction of America's resource base cannot be tolerated."

His warning was underlined by reports of heavy soil losses in areas where fencerow-to-fencerow planting was encouraged, including one by State Conservationist Wilson Moon showing that up to two bushels of topsoil is lost on much of Iowa's agricultural land for every bushel of corn produced. Farmers planting soybeans on unprotected, sloping land, he reported, can lose up to seven bushels of topsoil for every bushel of soybeans.

The Soil Conservation Service has been urging action for some time on a program to identify and preserve good agricultural land. The Department of Agriculture, in its first public response to SCS and other prodding, held a prime-land seminar last summer that brought eighty land experts together to chart a new federal course.

Their preliminary report conceded that cropland conversion has become a serious problem, particularly in New Jersey and several other urbanized states. It predicted that demand for food, fiber, and timber will eventually test U.S. production capability "although it is not certain when, or with what degree of urgency, this will occur."

The participants agreed that increasing pressure on cropland involves resource utilization conflicts, particularly between agriculture and energy development, as well as regular growth in demand for agricultural products both at home and abroad.

"In some areas competition for limited water resources may be as important as competition for land," the experts concluded. "Utilization of water for energy production may preclude irrigation on

vast acreages of potentially productive land, resulting in increased demand elsewhere to meet production needs."

It is clear that putting good cropland under
concrete, or converting it to other nonagricultural
uses, will continue to force food production onto
more marginal land. It is inevitable that this will
increase production costs and make food cost more,
particularly in view of population increases
projected both in the United States and elsewhere in
the world.

It also is clear that these agricultural land losses
will continue to close out opportunities for vegetable
and fruit production units near urban areas, reducing the opportunity for land application of city
wastes and for people to purchase more fresh,
locally grown food.

Finally, as the Citizens Advisory Committee on
Environmental Quality report pointed out: "It is
critical that cities retain the remaining alluvial
valleys and other fertile land in agricultural use, lest
we wake up some day to continuous development,
clogged highways, commercial strips, and an empty
breadbasket."

Roger Blobaum

Index

Yields *(cont.)*
    Corn Belt study, 95–97
    interplanting and, 116–117, 118–121
    manure and, 179
    mulched potatoes and, 126–127
    sludge and, 198

Zinc, 4, 61
    cadmium and, 195
    in sewage sludge, 191–193